VIBROACOUSTIC CONDITION MONITORING

ELLIS HORWOOD SERIES IN MECHANICAL ENGINEERING

Series Editor: J. M. ALEXANDER, formerly Stocker Visiting Professor of Engineering and Technology, Ohio University, Athens, USA, and Professor of Applied Mechanics, Imperial College of Science and Technology, University of London, and Professor S. REID, Department of Mechanical Engineering, University of Manchester Institute of Science and Technology

The series has two objectives: of satisfying the requirements of postgraduate and mid-career engineers, and of providing clear and modern texts for more basic undergraduate topics. It is also the intention to include English translations of outstanding texts from other languages, introducing works of international merit. Ideas for enlarging the series are always welcomed.

Alexander, J.M.	**Strength of Materials: Vol. 1: Fundamentals; Vol. 2: Applications**
Alexander, J.M., Brewer, R.C. & Rowe, G.	**Manufacturing Technology Volume 1: Engineering Materials**
Alexander, J.M., Brewer, R.C. & Rowe, G.	**Manufacturing Technology Volume 2: Engineering Processes**
Atkins A.G. & Mai, Y.W.	**Elastic and Plastic Fracture**
Beards, C.F.	**Vibration Analysis and Control System Dynamics**
Beards, C.F.	**Structural Vibration Analysis**
Beards, C.F.	**Noise Control**
Beards, C.F.	**Vibrations and Control Systems**
Borkowski, J. and Szymanski, A.	**Uses of Abrasives and Abrasive Tools**
Brook, R. and Howard, I.C.	**Introductory Fracture Mechanics**
Cameron, A.	**Basic Lubrication Theory, 3rd Edition**
Cempel, C.	**Vibroacoustic Condition Monitoring**
Collar, A.R. & Simpson, A.	**Matrices and Engineering Dynamics**
Cookson, R.A. & El-Zafrany, A.	**Finite Element Techniques for Engineering Analysis**
Cookson, R.A. & El-Zafrany, A.	**Techniques of the Boundary Element Method**
Edmunds, H.G.	**Mechanical Foundations of Engineering Science**
Fenner, D.N.	**Engineering Stress Analysis**
Fenner, R.T.	**Engineering Elasticity**
Ford, Sir Hugh, FRS, & Alexander, J.M.	**Advanced Mechanics of Materials, 2nd Edition**
Gallagher, C.C. & Knight, W.A.	**Group Technology Production Methods in Manufacture**
Gohar, R.	**Elastohydrodynamics**
Haddad, S.D. & Watson, N.	**Principles and Performance in Diesel Engineering**
Haddad, S.D. & Watson, N.	**Design and Applications in Diesel Engineering**
Haddad, S.D.	**Advanced Diesel Engineering and Operation**
Hunt, S.E.	**Nuclear Physics for Engineers and Scientists**
Irons, B.M. & Ahmad, S.	**Techniques of Finite Elements**
Irons, B.M. & Shrive, N.G.	**Finite Element Primer**
Johnson, W. & Mellor, P.B.	**Engineering Plasticity**
Kleiber, M.	**Incremental Finite Element Modelling in Non-linear Solid Mechanics**
Kleiber, M. & Breitkopf, P.	**Finite Element Methods in Structural Engineering: Turbo Pascal Programs for Microcomputers**
Leech, D.J. & Turner, B.T.	**Engineering Design for Profit**
Leech, D.J. & Turner, B.T.	**Project Management for Profit**
Lewins, J.D.	**Engineering Thermodynamics**
Malkin, S.	**Materials Grinding: Theory and Applications of Machining with Abrasives**
Maltback, J.C.	**Dynamics in Engineering**
McCloy, D. & Martin, H.R.	**Control of Fluid Power: Analysis and Design, 2nd (Revised) Edition**
Osyczka, A.	**Multicriterion Optimisation in Engineering**
Oxley, P.L.B.	**The Mechanics of Machining**
Piszcek, K. and Niziol, J.	**Random Vibration of Mechanical Systems**
Polanski, S.	**Bulk Containers: Design and Engineering of Surfaces and Shapes**
Prentis, J.M.	**Dynamics of Mechanical Systems, 2nd Edition**
Renton, J.D.	**Applied Elasticity**
Richards, T.H.	**Energy Methods in Vibration Analysis**
Ross, C.T.F.	**Computational Methods in Structural and Continuum Mechanics**
Ross, C.T.F.	**Finite Element Programs for Axisymmetric Problems in Engineering**
Ross, C.T.F.	**Finite Element Methods in Structural Mechanics**
Ross, C.T.F.	**Applied Stress Analysis**
Ross, C.T.F.	**Advanced Applied Stress Analysis**
Ross, C.T.F.	**Finite Element Methods in Engineering Science**
Roy, D. N.	**Applied Fluid Mechanics**
Roznowski, T.	**Moving Heat Sources in Thermoelasticity**
Sawczuk, A.	**Mechanics and Plasticity of Structures**
Sherwin, K.	**Engineering Design for Performance**
Stupnicki, J.	**Stress Measurement by Photoelastic Coating**
Szymanski, A. and Borkowski, J.	**Technology of Abrasives and Abrasive Tools**
Szczepinski, W. & Szlagowski, J.	**Plastic Design of Complex Shape Structured Elements**
Thring, M.W.	**Robots and Telechirs**
Walshaw, A.C.	**Mechanical Vibrations with Applications**
Williams, J.G.	**Fracture Mechanics of Polymers**
Williams, J.G.	**Stress Analysis of Polymers 2nd (Revised) Edition**

ELLIS HORWOOD SERIES IN AUTOMATED MANUFACTURING

Series Editor: COLIN BESANT, Professor of Computer-Aided Manufacture, Imperial College of Science and Technology, London University

Besant, C.B. & Lui, C.W.K.	**Computer-aided Design and Manufacture, 3rd Edition**
Ding, Q.L. & Davies, B.J.	**Surface Engineering Geometry for Computer-aided Design and Manufacture**
Gosman, B.E., Launder, A.D. & Reece, G.	**Computer-aided Engineering: Heat Transfer and Fluid Flow**
Gunasekera, J.S.	**CAD/CAM of Dies**
Vickers, G.W., Ly, M. & Oetter, R.G.	**Numerically Controlled Machine Tools**

VIBROACOUSTIC CONDITION MONITORING

C. CEMPEL Ph.D., D.Sc.
Head of Applied Mechanics Institute
Technical University of Poznan, Warsaw, Poland

Translation Editor
SAM D. HADDAD
Consulting Professor, Stanford University, USA

ELLIS HORWOOD
NEW YORK LONDON TORONTO SYDNEY TOKYO SINGAPORE

First published in 1991 by
ELLIS HORWOOD LIMITED
Market Cross House, Cooper Street,
Chichester, West Sussex, PO19 1EB, England

A division of
Simon & Schuster International Group
A Paramount Communications Company

Printed and bound in Great Britain
by Boocraft Ltd, Midsomer Norton, Avon

British Library Cataloguing-in-Publication Data

Cempel, Czesław
Vibracoustic condition monitoring —
(Ellis Horwood series in mechanical engineering)
I. Title. II. Series
620.3
ISBN 0–13–931718–X

Library of Congress Cataloging-in-Publication Data

Cempel, Czesław.
Vibroacoustic condition monitoring / C. Cempel: translation editor, Sam D. Haddad.
p. cm. — (Ellis Horwood series in mechanical engineering)
Translated from Polish.
Includes bibliographical references and index.
ISBN 0–13–931718–X
1. Machinery — Monitoring. 2. Acoustic emission testing. I. Haddad. S. D. (Sam David), 1939– . II. Title. III. Series.
TJ153.C39 1991
621.8′16–dc20 91–24791
CIP

Table of contents

Preface

Since I published *Bases of vibroacoustical diagnostics of machines*, many changes have taken place in the state of knowledge and achievements in vibroacoustical diagnostics of machines. Also, the range of diagnostic needs has changed requiring corresponding changes in the field of education. Therefore, instead of writing a monograph where simple problems can be omitted, it became necessary to write a book on vibrational diagnostics, with emphasis on the presentation of a complete and coherent picture of the method rather than the reporting of most recent cognitive and application achievements in this field. The present book intends, therefore, to fill the gap which refers to the diagnostics of machines and installations in general.

CONTENT ARRANGEMENT

The first chapter presents the origin of the new field against a background of diagnostics and nondestructive tests of machines in general, as well as the possibilities to utilize diagnostics in the design, manufacturing and use of machines and installations. The chapter also presents results of the most recent investigations with reference to reliability and cost effectiveness as related to the implementation of vibroacoustical diagnostics in industry.

The second chapter presents physical premises motivating the application of vibration, noise and acoustics emission in diagnostics, and clarifies the possible qualitative relations between the wear processes and the vibroacoustical processes. From this we derive the notion of tribovibroacoustical (wear-vibroacoustical) processes. Based upon this notion, the majority of failures in machine elements (e.g. rotors, gears and bearings) as well as a number of the basic wear processes (e.g. fatigue, friction, creep and cavitation) are discussed.

The third chapter presents models and methods of diagnostic observation of mechanical objects with the help of vibroacoustical processes. It introduces the basic paradigm of vibroacoustical diagnostics in the form of a cause-and-effect sequence: REAL CONDITION—VIBROACOUSTICAL PROCESS—SIGNAL—SYMPTOM—CONDITION ASSESSMENT. It gives quantitative and qualitative motivations in reference to selecting the proper process for diagnostic observation of the object, as well as presenting the possible methods of treatment in order to obtain vibroacoustical symptoms–possibly interchangeable with the condition of the object.

The fourth chapter presents the path which should be followed in order to assess present and future condition of machines as soon as vibroacoustical measurement is obtained. It presents the possible diagnostic models, i.e. the assignment of the 'symptom–condition' relation to mechanical objects, and the methods of inferring the proper condition for each model type. It also shows methods of minimization of disturbances in vibroacoustical symptoms and, hence, the possibility of increasing the reliability of the diagnostic decisions. Finally, the conclusion of the chapter presents possible methods of prognosis of the condition of mechanical objects, i.e. *the prognosis of the date and range of machine overhauls* during use.

The fifth chapter is a methodical exemplification of the proceeding chapters, and presents examples of diagnosing high-power fans, diesel engines, textile machines and electric motors.

Finally, the sixth chapter shows methods and examples of the discrimination of fault characteristics in a passive diagnostic experiment, and the way of distinguishing the condition. It also outlines new methods of adaptational and diagnostic filtration of machine fault signals.

The book is concluded with appendices containing a collection of the most important diagnostic standards and criteria of balancing, as well as a synthesis of some vibration diagnostics problems.

PROSPECTIVE READERS AND THE METHOD OF USING THIS BOOK

Based on the presented contents outline, we can expect a wide range of readers interested in the material contained in this book. The first three chapters can serve as supplementary material for students of mechanical vibrations, or vibroacoustical of mechanics. Postgraduate students are recommended to use the first five chapters in a compact form. The compact form is particularly important and has been verified by the Author during the production of this book, as well as during his lectures in Postgraduate Study of Diagnostics in 1984.

Diagnostic services can be found in large industrial plants. The members of such diagnostic staff will find material of interest to them in Chapters 1, 2, 3 and 5. On the other hand, the first chapter is particularly designed for production and maintenance managers in industry, since it gives economic premises for the appointment of a diagnostic group of specialists engaged in maintaining continuous operation of machines, or in the inspection of quality. Finally, Chapters 1 and 5 can form the basis for organizing short-term diagnostic courses for members of a maintenance organization, or courses carried out directly in industrial plants.

ACKNOWLEDGEMENTS

Words of gratitude are due to my colleagues from the Dynamics and Vibroacoustics of Machines Group involved in scientific and application work in the field of diagnostics. In particular, I would like to acknowledge Dr Franciszek Sordyl, Dr Marek Majewski, as well as Mr Maciej Andrzejewski, and Dr Ryszard Nowicki. A discussion with them on the problems of diagnostics during the work on the government program PR-8 in the plants ZA-Tarnów and Petrochemia Plock, contributed significant stimulating factors and sources of the research resulting in this book.
Separate thanks are expresses to Mrs Maria Zwierzykowska for her speedy and careful typing of the manuscript, and to Mrs Ewa Kandula who prepared very carefully the illustrations contributing greatly to an easier understanding of the presented ideas. Words of gratitude are directed also to Mrs Bozena Rakowska, the librarian of the Institute, who did her best in finding every needed literature item.

Expressing my warmest thanks to all those mentioned, as well as the not mentioned co-workers and contributors, I want to apologize for any possible shortcomings in the book for which I take sole responsibility.

C. Cempel

1

Introduction

1.1 ORIGIN OF MACHINERY DIAGNOSTICS AND VIBROACOUSTICAL DIAGNOSTICS

In recent years technology has been characterized by an increased complexity on the one hand, and an increased responsibility of tasks on the other. Hence, we witness the creation and development of a new field of science and technology: the *technical diagnostics* which developed in response to the demand by the users of these sophisticated technical tools. The purpose of the new field is the determination of the widely understood technical condition of installations with the help of objective methods and means in order to increase their life, reliability and efficiency of operation. Such knowledge of the *technical condition* of a machine or installation, determined by a set of characteristics, is necessary for taking decisions in each phase of machine life (i.e. in the phases of design and investigation of prototypes, machine manufacturing and in its use) based on a good understanding of its condition.

A wider significance of diagnostics consists according to Ziemski [1] and Pszczołowski [2] in the recognition of the given object and its developmental tendencies on the basis of its symptoms, and the knowledge of general regularities for the purpose of a planned action. This definition is wide enough to include the health diagnosis of a human individual, the technical condition of a machine, installation, or the diagnosis of the state of the organization and management of an enterprise. Our subject of interest is the technical diagnostics of machinery and installation, beginning with the observable vibrations and noises, i.e. the vibroacoustical symptoms of the operation and condition of these objects.

The need to diagnose the condition of equipment was first perceived in electronic engineering relating to manufacturing and operation of computers. From there, the development of this branch of science has led to the invention of apparatus and self-diagnosing procedures indicating the place and type of fault in the complex computer system.

In mechanical engineering, in construction (or design), manufacturing and operation of machines and installations, the factor which directly stimulates the development of diagnostics is the ability to fulfil the intended function or mission. This responsibility can be defined in the categories of human safety or economic productivity and effectiveness. It is easy to guess that diagnostics of machines and mechanical installations was first developed and applied in air and sea transport, and in power engineering. At present, we witness the adaptation and development of new methods and means for the assessment of the state in the remaining branches of industry and national economy. There are three factors which have prompted the development of diagnostics. The first is the complexity of production systems where the breakdown of one machine stops the whole process line giving in effect expressive economic losses. The second, particularly true for small and medium machines, is the great number of these machines operating simultaneously in a continuous cycle and without any supervision. For example, an average refinery or chemical plant uses several thousand small and medium electric direct drive motors simultaneously. The maintenance and overhaul of such numerous machine stock causes many problems when we are not able to predict the date of overhaul. Finally, the third factor consists in the high level of reliability required of some installations for single or seasonal use. This reliability is required of special vehicles such as spaceships and military and agricultural equipment where, for example, one machine waits the whole year to be used for several weeks during harvesting, or other annual events.

1.1.1 Machine diagnostics

It seems there are no major difficulties in transferring diagnostic experiences from highly developed computer techniques to the field of machine diagnostics. Computer electronic elements, each with as high degree of repeatability, operate in a discrete and determined way. Hence, the sequence of the functional conditions reflecting the correct operation of the definite subsystem can always be predicted. On the other hand, the process of machine wear has a continuous character and gives no premises for a natural separation of the conditions. Furthermore, mechanical elements and machines are characterized by a greater dispersion of characteristics acquired in their stage of manufacturing than is the case with electronic elements. This indeterminacy and the continuity of the wear process creates different problems in the control of the characteristics and the condition of mechanical installations. Therefore, the diagnostics of machines having similar objectives and tasks must develop a different set of methods and means for diagnosis. Having this in mind, it is now possible to define the utilitarian objective of machine diagnostics. it can be characterized by the following tasks included, according to Pszczolowski [2], in the notion of complete diagnostics.

(1) *Diagnosis*—determination of the current technical condition.
(2) *Origin*—determination of the causes of the present condition.
(3) *Prognosis*—determination of the lead time for a future change in the technical condition.

In this context, the role of science in diagnostics is reduced to supplying reliable methods and means for the realization of the above tasks. Premises for an objective assessment of the condition of machines are provided by tests. So, from the executive point of view, diagnostics consists of the measurement of the observable symptoms of the technical condition, and inference on the basis of obtained sets of data according to a model of the analysed object (machine, installation). The possibilities of such tests and diagnostic observations can easily be perceived when we look at the machine as an operating system with energy and information flow, as it is shown in Fig. 1.1.

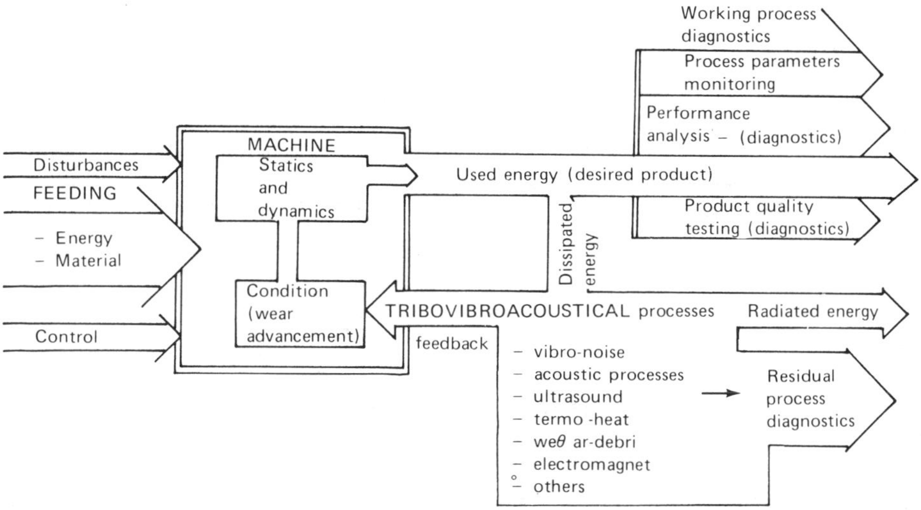

Fig. 1.1 — The machine as an operating system with energy and information flow and the possibility of its diagnostic observation

The first input of the machine system (i.e. the supply of energy and raw material) is responsible for the flow of energy and the action of forces in the machine. From the point of view of diagnostics, the state of this input is essential but such information is rarely directly available. Similarly, the input of control determining the way a machine works, is not previously known in the majority of cases. The same refers to the input of disturbances created in the cooperation of the machine with its surroundings. Therefore, in the majority of diagnostic cases, in the procedure of defining the technical condition by observation (measurement) of different conditional symptoms, we must be aware of the fact that *all inputs of the machine system are unknown.*

The notion of the *measurable symptom of the condition* is very broad, and incorporates three groups of characteristics or parameters accessible for diagnostic observation. In the first, through diagnostics, we can investigate the *functional parameters* (working parameters of the machine, such as power, maximum velocity, etc. This method of diagnostics requires a periodical shutdown of the machine and subjects it to a determined sequence of special tests. The second type is the measurement of *parameters and characteristics* being the *direct symptoms of wear.* This is primarily concerned with the examination of deviations in the form and dimensions in relation to the base line of a machine in good condition. This type of examination is usually associated with a periodical overhaul, since it requires not only a shutdown of the machine but also its dismantling.

The last and most promising method is the examination of *residual processes* which in an unintended but inseparable way to accompany the functioning of machines. These processes are illustrated in detail in Fig. 1.2. Their energy effect (in the understanding of

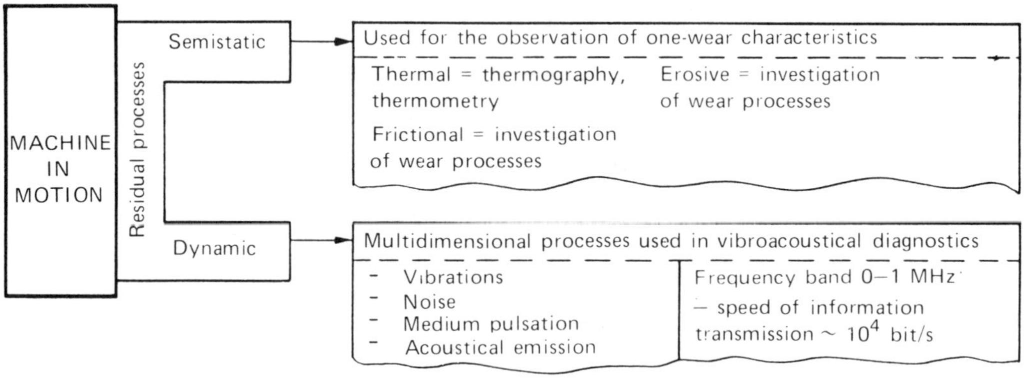

Fig. 1.2 — Diagnostic specification of residual processes accompanying the function of machines

energy supplied to the machine) is minimal, 10^{-6} and less, but their feedback, in the majority of cases, has a *destructive damaging* character (Fig. 1.1). These are thermal, frictional, electric and primarily vibroacoustical processes in the form of vibrations, noises and pulsations of the working medium in the machine. In many cases, these processes are symptoms of wear (e.g. products of abrasion) or the determinants of wear, such as vibrations causing fatigue of material. Their utilization in diagnostics creates a chance of *assessing the technical condition without dismantling* the machine, without shutting it down, and even *without a direct contact,* as in the case of noise measurements and vibration measurements by lasers, etc.

1.1.2 Vibroacoustical diagnostics

The possibility of assessing the technical condition of a machine during normal working by observation of vibroacoustical processes which bear considerable information content (Fig. 1.2) was very quickly recognized in aircraft technology and in power engineering (the vibroacoustical process = dynamic process taking place in the machine or its surroundings, i.e. the vibrations, noise, pulsation of medium, acoustical emissions occurring within the frequency range of 0–1MHz and more). This became possible, among other reasons, when electronics created enormous possibilities of measurement and analysis of vibroacoustical signals. Because of this we are now able to select an optional component of the signal and to transform it into a symptom interchangeable with the technical condition of the analysed element or subassembly in the complex mechanical object, e.g. rolling element bearing in a reaction motor. Furthermore, form the same vibration signals, we can obtain symptoms indicating an advancement of wear for independent ways of damaging, e.g. unbalance of the rotor, condition of rolling bearing, presence of cavitation, etc. At present vibroacoustical diagnostics give possible methodical and instrumental; answers to the following questions.

(1) What is the condition of the prototype of the machine (element subassembly)?—*Construction or design diagnostics.*
(2) What is the condition of the machine (element, assembly) after its manufacture?—*Checkout diagnostics.*
(3) What is the condition of the machine of the machine (element, assembly) during its use? — *Condition monitoring.*
(4) How is the production process realized by the machine?—*Diagnostics of production processes.*

We will answer these questions by observing the vibroacoustical processes accompanying the operation of machines using adequate means and methods. Thus, the role of science consists here in the development of objective methods and means of observation and diagnostic inference, and in the presentation of the final diagnostic algorithm (procedure) being a sequential answer to the following questions.

(1) What and where should be measured? (Which process? In which place and direction?)
(2) How to measure? (What parameter, symptom? How often?)
(3) How to infer? (Limit conditions, classes of conditions, preventative measures.)

Since, in the majority of diagnostics cases, we do not know the conditions of the input of the machine system, therefore our diagnostic algorithm cannot be sensitive to changes in the input power, control and disturbances in the machine (Fig. 1.1).

The purpose of this book is to attempt to answer the questions formulated above and, since in many cases such answers have not been given yet, the objective here is to present methods helping to obtain the answers.

1.2 POSSIBILITIES OF MACHINE CONDITION TESTING AND VIBROACOUSTICAL DIAGNOSTICS

Before we proceed to investigate the basis of vibroacoustical diagnostics, we should review all possible testing methods of the conditions of machines, construction and machine elements. There are many such methods and they can be divided into two categories (Table 1.1). The first group comprises methods which require a special excitation for obtaining the assessment, such as a light source, ultrasound wave, magnetic field, Roentgen rays, etc. These methods are called nondestructive testing. However, they generally can be used only for separate elements of a machine or construction.

The second group of methods based on the observation of residual processes accompanying the functioning of machines can be called the passive diagnostic tests. The simplest type of these diagnostic methods is analysis of wear products contained in the lubricating oils and hydraulic oils. Many well-mastered detailed techniques exist, (for details see, for example, Collacott [3] but among them only the spectrometric oil analysis) distinguishes the type of damage using very expensive laboratory equipment.

Thermal diagnostics utilizing thermovisual cameras is especially useful in the detection of faults in thermal systems. On the other hand, we apply the measurements of

Table 1.1 — Some important testing methods of machines, constructions and their elements [3, 4]

Name of method	Essentials	Area of application	Limitations
I. STIMULATING METHODS (nondestructive tests)			
Visual inspection			
— endoscopic	Visual inspection through a set of lenses or lightpipe fibre	Immovable elements of machines and constructions	Only superficial faults; direct access necessary
— holographic	Reconstruction of the wave front from 3-D diffraction pattern	Immovable elements of machines and constructions	Only superficial faults; direct access necessary
— penetrating, colour or fluorescent	Penetration into the faults by visible or fluorescent chemicals	Immovable elements of machines and constructions	Only superficial faults; direct access necessary, faults ranging within millimetres
Magnetic methods			
— powder	Concentration of ferropowder around the faults and failures	Immovable elements of machines and constructions of ferro materials	Only superficial faults; direct access necessary, faults ranging within millimetres
— rotocurrent	Change of amplitude and current phase around the fault	Immovable elements of machines and constructions, but of conductive materials	Only superficial faults; direct access necessary, faults ranging within millimetres
Radiography			
— X-ray — isotopic — neutron	Damping, reflection, dispersion of penetration radiation or neutron flux	Immovable elements of machines and constructions	Expensive and heavy instrumentation, direct access necessary. Volumetric faults; Minimum fault size 2–4 mm. The smallest fault size in the range of millimetres
Ultrasound	Damping, dispersion, reflection, phase change, resonance change of incident	Immovable elements of machines and constructions	Direct access necessary

Name of method	Essentials	Area of application	Limitations
Acoustical emission	Forced development of microdamage (e.g. by stress field) is the source of acoustical emission of transient waves with high frequency	Immovable elements of machines and constructions	Expensive instrumentation, limited experience
II. DIAGNOSTIC METHODS (passive diagnostic tests)			
Testing of wear products			
— filter elements	Lubricating oil (filtered, centrifuged working liquid) gives wear products for quantitative examination	Machines in motion, after sampling of oil (liquid)	Difficulties in failure localization
— magnetic corks	Interception of big magnetic particles in oil (working liquid)	Machines in motion, after sampling of oil (liquid), after replacement of cork	Difficulties in failure localization, only ferromaterials, big particles
— ferrography	Interception of big magnetic particles in oil (working liquid), but all particles	Machines in motion, after sampling of oil (liquid)	Difficulties in failure localization, only ferromaterials, big particles
— spectral oil analysis SOA	Spectrographic analysis of small particles after burning	Machines in motion, after sampling of oil (liquid), possibility to distinguish failure types	Expensive instrumentation
— counting of particles	Current counting of particles lubrication system (hydraulics)	Machines in motion, in the current readout	Expensive and troublesome installation
Thermic diagnostics			
— thermography	Analysis of object image in infra-red	Unmovable machine elements and installations subject to getting hot	Direct access necessary

Table 1.1 — Some important testing methods of machines, constructions and their elements [3, 4] continued

Name of method	Essentials	Area of application	Limitations
II. DIAGNOSTIC METHODS (passive diagnostic tests) continued			
Thermic diagnostics			
— thermometry	Point measurement of temperature by thermocouple or the like	Slide bearings, bodies of power machines	Difficulties in the installation, great inertia
Vibroacoustical diagnostics			
— vibrations	Analysis of vibrations connected with the functioning of machine (subassembly)	Machines (subassemblies) in motion, possibility of contactless examination	NO SIGNIFICANT LIMITATIONS
— noise	Analysis of vibrations connected with the functioning of machine (subassembly), noise analysis	Machines (subassemblies) in motion, possibility of contactless examination	Sensitivity to external disturbances
— pulsation of medium	Analysis of medium pulsation in working areas of machines	Subassemblies of fluid-flow machines (internal combustion engines, pumps, etc.)	Difficult access to the measurement area
— acoustical emission	Reception and analysis of acoustic waves in the material generated by the natural development of micro damage	Elements of machines (constructions) under working load	Expensive instrumentation

he temperature of slide bearings, and the bodies of power machines with the help of permanently mounted thermocouples, etc. From the diagnostic point of view, the temperature of an element is not a good carrier of information concerning the condition of machines, since it refers to the mounting point of the thermocouple, while the rate of temperature reaction to changes in supervised condition of the element is not great. The problem is different in reference to vibroacoustical phenomena comprising vibration, noise, pulsation of the working medium and acoustical emission induced by the working load. As mentioned before (particularly in the observation of machine vibrations on the basis of one process we can assess the condition of the whole machine and its main subassemblies since the speed of transferred information and its variety is very great (Fig. 1.2). The vibroacoustical diagnostics—the only method of machine evaluation without dismantling and carried out during machine operation—has one essential difficulty which is the lack of a *code which could translate the technical condition* of a machine into the generated vibroacoustical process. The breaking of this code, or the *extraction of the original signals and vibroacoustical symptoms of wear* is one of the main tasks of machine vibroacoustical diagnostics.

The above review of the testing methods of a machine's condition, construction and elements, is methodically presented in greater detail in Table 1.1. A similar survey, but concentrated on machine diagnostics, is shown in Fig. 1.3, where the methods of vibroacoustical diagnostics are shown in capital letters. As we can see from the figure, each element of machine construction is treated separately or it can be investigated in the operating machine with the help of vibroacoustical diagnostics or ultrasound nondestructive methods. So, in the application of these methods there is a great potential of an objective inference about machine condition.

1.3 THE ROLE OF VIBROACOUSTICAL DIAGNOSTICS IN MACHINE LIFE

Machines, installations, as well as any other product pass through three phases or stages: (a) design and manufacturing; (b) use or utilization; and (c) renovation or overhaul. The transference from one phase into the next requires some objective methods and some means of assessing the quality of the new product or its technical condition in the stage of utilization. These methods can be provided by vibroacoustical diagnostics.

1.3.1 Design diagnostics

According to Dietrych [5], design means the selection of construction characteristics permitting an optimal realization of the basic task of a machine. One of the measures of the ability to realize the tasks (objective function) are the indices of quality to which can be included expected life, reliability, precision, and low level of emitted external disturbances (vibrations and noise). These indices are inseparably connected with the character of the dynamic process and also of the vibroacoustical processes taking place in machines [6]. An illustration of the causal relationship between these quality indices and vibrations are presented in Fig. 1.4.

It is obvious that too much vibrational activity of the machine (dynamic deformations, stresses) exert a negative effect on the durability, precision and reliability of the product. Hence, in the stage of design and examination of the prototype, it is

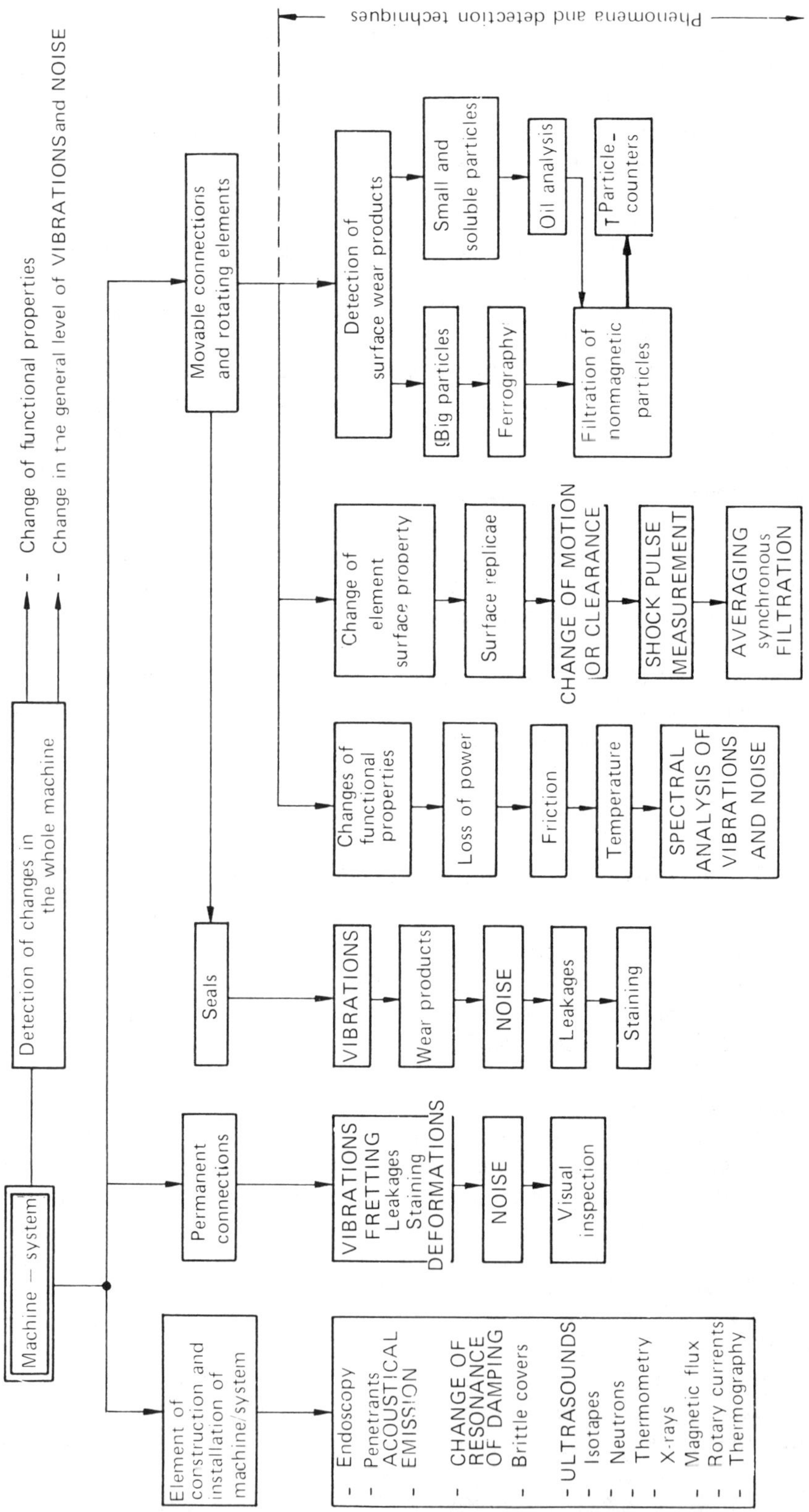

Fig. 1.3 — Possibilities of detecting wear of machines and their parts with the help of technical diagnostics and nondestructive tests—methods of vibroacoustical diagnostics are marked by capital letters [7]

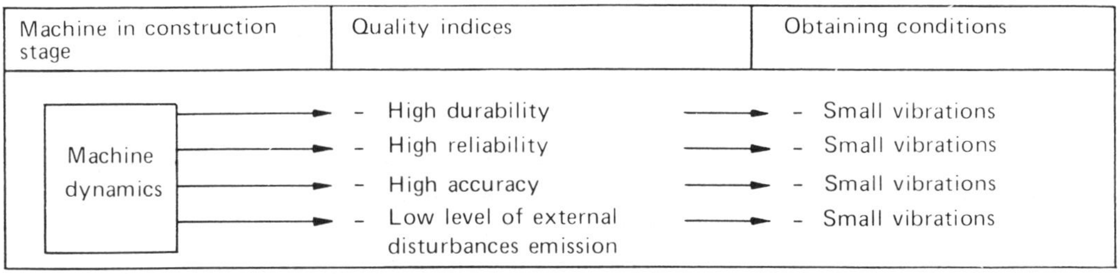

Fig. 1.4 — Indices of machine quality and their connection with vibrations

necessary to localize the sources of vibrations and noise. The information obtained about the localization, relative power and origin of the vibration source permits subsequent change to the construction in order to reduce the vibration activity of the machine to levels admissable by the corresponding standards or other determinations. Such vibroacoustical examination of machines aiming at the localization of the sources and the determination of the character of their generation and mutual intensity is called vibroacoustical design diagnostics [7].

1.3.2 Checkout diagnostics (quality control)

Positive results of the above diagnostic examinations permit shifting from the design phase to the manufacturing phase of the machine or other technical product. The manufacture of machine elements and mechanisms is inseparably connected with the addition to the determined project of two types of random components. Firstly, within the given class of working tolerance there is always a *random spread of dimensions* between different items of the same products. Secondly, a similar *random* component can be observed by measuring the *mechanical properties* of these items or, rather, the materials of which the items have been made. The categories of macroscopic properties refer to elasticity, plasticity, hardness, impact resistance, vibration damping, etc. In the field of microscopic properties, it is important to note the incidental differences in the physiochemical composition and the microstructure of machine elements, and particularly their surface layers, since they are most exposed to the action of the complex load in the later process of utilization.

The next stage of manufacturing is the assembly of the elements into one functioning unit. Also here, in spite of a careful selection and pairing of rotational and sliding pairs and permanent connections manufactured by different methods (welding, gluing, riveting, screw joints, keys, etc.), each stage brings unavoidable incidental deviations from the standard of an ideal assembly, e.g. differences in the tension of screw joints, in the length and thickness of weld, glue, etc. In the light of this, it is clear that the analysis of the products of the same type will show a random distribution of their geometrical and mechanical properties around a mean-determined property of the standard. The dynamic processes taking place during the operation of machines will be different, and the degree of their mutual differentiation will depend on the *history of randomness* in the whole stage of manufacturing. The difference in the geometrical and mechanical properties and the resulting spread of dynamic processes of the particular machines will be revealed finally in the spread of their *functional properties*. Therefore, there comes into play the *checkout diagnostics* as the final link in the manufacturing process qualifying the products as *good or faulty* for use. Examples of such standard procedures are vibrational control of the balancing of new turbomachines, inspection of vibrations and noise of new bearings, electric motors, etc.

1.3.3 Condition monitoring

The next stage of machine life is its utilization. Here, the main effort of constructors, manufacturers and users is the reliable, smooth and efficient operation of the machine. The standard time of such operation is determined and guaranteed by the producer on condition that the technical and operational instructions, as well as the overhaul and repair dates are obeyed. In spite of that, in the utilization process there occur two diametrically different facts. Firstly, there exists a group of machines for which the smooth operation time is considerably shorter than the standard time. Secondly, overhauls carried out in the standard time of the given group of machines do not find any justification in one part of the machines since they are in good technical condition, and at the same time such overhaul usually causes their condition to deteriorate. There are four causes of such great random spread in the smooth operation time: (a) the earlier described random *differentiation of properties* obtained in the manufacturing stage; (b) *differentiation in the working load*; (c) *different reaction with the surrounding* (assembly of connections, foundation) among the particular machines, and (d) quality differences in successful repair operations.

Objective necessities of safety and economics force the users of machines, on the one hand, to minimize unplanned overhauls and breakdowns, and, on the other, to eliminate unneeded overhauls of machines in good technical condition (in spite of having exceeded the standard time of operation). When we uphold the assumption that the dynamic processes in a machine reflect its utilization condition, then a remedy for the solution of the majority of the above-mentioned tasks, without shutting down the machine, is the vibroacoustical diagnostics. These particular purposes will be served by *vibroacoustical condition monitoring*. Depending on the definite needs, this type of diagnostics can comprise only the determination of the degree of the object's fitness for further utilization, i.e. a *two-condition diagnosis:* good/faulty. A further possibility is the determination of the type and cause of failure (i.e. the origin) and the determination of the time horizon of the future change of condition, or *prognosis* (prediction).

One of the elements of rational utilization of machines is their renovation or regeneration which may consist in the adjustment, repair or replacement of elements or subassemblies [8]. On the basis of information obtained from diagnostic observation of a machine, we can plan the date and extent of the regeneration procedures together with the preparation of materials and subassemblies necessary for replacement. Furthermore, diagnostic examinations carried out immediately after overhaul, permit the determination of the *quality of the regeneration procedures,* and the avoidance of starting the machine before it is completely efficient.

The use of vibroacoustical diagnosis or condition monitoring of machines under continuous running can assume different observation forms. For unique objects with vital tasks a *permanent monitoring* is installed with multipoint measurements of vibrations and other vibroacoustical processes. For machines with less critical tasks, a one-point measurement of vibrations carried out periodically intervals (each of several weeks) is sufficient. This method of observation can be called the *diagnostic inspection*

1.3.4 Diagnostics of production processes

Machines and installations participate in complex production processes with an increasing demand for automatic control without the interference of human operators. At the same time, we know that particular phases of the given production process differ by a different character and intensity of the emitted vibroacoustical processes. Therefore, the opportunity arises to utilize these phenomena for the checkout and control of the production processes. Such an approach to the problem creates the possibility of automation of the process and it gives an additional increase of the effectiveness and quality of the process. This type of vibroacoustical diagnostics application in industry is called the *vibroacoustical diagnostics of the production processes.*

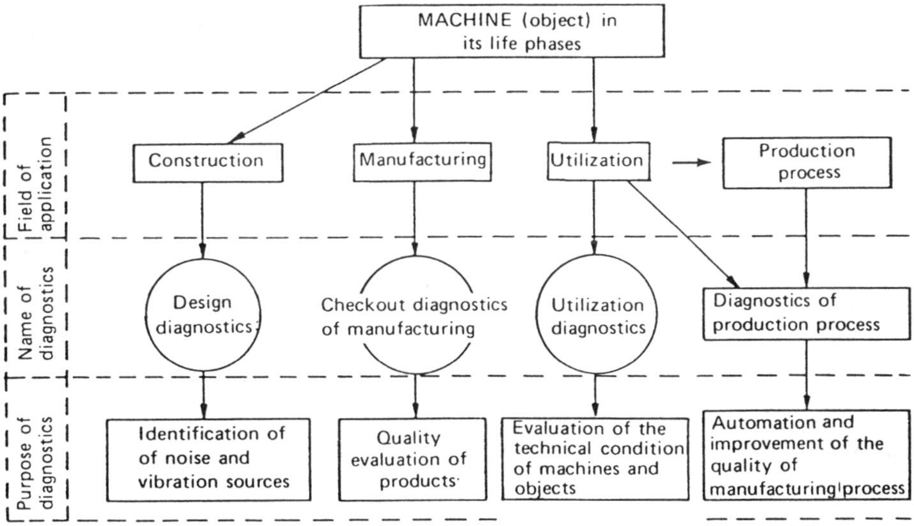

Fig. 1.5 — A synthetic presentation of the objectives of the particular types of vibroacoustical diagnostics in industry [7]

The above-mentioned applications of vibroacoustical diagnostics in industry are artificially presented in Fig. 1.5, showing the great range of information which can be obtained from the examination of vibroacoustical processes generated by machines. The methods of extracting information from the vibroacoustical signal (process) are the subject of vibroacoustical diagnostic investigations and we shall discuss them in the following chapters. The next section of this chapter is devoted to the economic and reliability consequences of the utilization of vibroacoustical diagnostics in industry.

1.4 ECONOMIC AND RELIABILITY MOTIVATION OF VIBROACOUSTICAL DIAGNOSTICS

As stated at the beginning of this chapter, vibroacoustical diagnostics was created as a branch of science and applied science in response to the growing demand of industry. In the fields of design, manufacture and utilization there exists a widely understood law of economic values and if only for this reason, it is worth elucidating this problem by the most spectacular example of the *utilization diagnostics* of machines.

Two methods of machine stock regeneration have existed in the field of machine utilization for many years: (a) overhaul after breakdown; and (b) planned preventive overhaul. The first method is used in cases when the breakdown and overhaul standstill do not cause any major losses. On the other hand, in the second method the planned overhaul of machines realizes the tasks it has to perform in order to prevent any unexpected breakdowns and the resulting losses. However, this type of machine utilization is characterized by two unfavourable features: (a) it does not ensure that unexpected breakdown will be avoided; and (b) the greatest number of breakdowns takes place immediately after preventive overhauls (see the example quoted below of a chemical plant in Norway). Therefore, for a group of machines realizing responsible tasks, a third method of utilization should be accepted in which the *overhaul* of a machine is *determined* only by its *technical condition*. It means that the dates, causes and range of overhaul are determined with the help of the methods and means of vibroacoustical diagnostics.

In the system of the after-breakdown overhauls significant effects also are obtained by partial diagnostics which supply information about the range of overhaul and the type of replacement parts required. In each of these systems, vibroacoustical diagnostics form an effective assessment method of the overhaul quality and supplies an *objective decision concerning the admission of the machine to utilization*.

The advantages derived from the application of vibroacoustical diagnostics in the utilization of machines are numerous and they can be divided into the following three categories.

(1) Increased safety.
(2) Increased productivity.
(3) Increased production quality.

A closer look at these advantages in qualitative understanding is possible in Table 1.2 [9].

The fields of industry in which these benefits bring the greatest profits include plants under continuous running in such branches as the following.

(1) Chemical industry.
(2) Mining and processing of coal, metals and crude oil.
(3) Production of electricity, gas and water.

The value of savings reaches the range of 0.5–3 per cent of pure production of the enterprise [9] depending on the economy sector.

Some numerical data for the year in which vibroacoustical diagnostics was introduced into a plant are quoted, among others by Hudachek [10] on the basis of his experiences in the refineries of Chevron Oil, USA. We can see from Fig. 1.6, the number of failures in the year decreased in one of the refineries from fifty to sixteen. In this plant, during a 7-year period after the introduction of vibroacoustical diagnostics, the following economic and reliability data were obtained:

(1) Decreased costs of repairs by 20–30 percent depending on machine type.

Table 1.2 — Benefits resulting from the introduction of vibroacoustical diagnostics in industry [9]

Benefits		Methods by which the benefits can be achieved: Failure prediction	Better knowledge of machines
Safety			
Decreased number of accidents caused by machines		Permits safe shutdown if immediate stoppage is not possible	Alarm of the supervising system is a sufficient reason for shutdown if an immediate stoppage is possible
Decreased threat for the health of personnel by vibrations and noise		Permits personnel to withdraw beyond the zone threatened by failure	Unnecessary expenditures for additional decrease of vibrations and noise are avoided
Productivity–Effectiveness			
Increased availability of machines	Longer time of utilization	Permits avoidance of losses owing to unplanned shutdown	
	Shorter time of overhaul	Permits operation to stop in time without failures and damage lengthening the time of overhaul. Permits the maintenance crew to start overhaul with already prepared replacement parts.	Decreases the time of machine condition testing after stoppage and accelerates the undertaking of adequate overhaul actions
Increased productivity		Permits productivity to be planned in time	Permits utilization of some machines with increased load or velocity. Permits detection of decreased efficiency of machines or increased energy consumption
Quality			
Better quality of products and services		Permits advance planning in order to avoid breaks in the supply to the buyer of products or services and increase thereby the good reputation of the producer	Permits decrease in the number of rejected products or reduction in the number or services not coming up to standard

(2) Increased time interval between overhauls from 3 to 7–8 years.
(3) Reduction of energy use by 1 percent.
(4) Reduction of nuisance and danger caused by vibrations and noise.

Even more optimistic data were quoted by a consulting company, VCI Ltd, from England, based on their industrial practice [11]. These data indicate that in 3000 machines, (Fig. 1.6(b)) the number of failures per month decreased fourfold during a 2.5-year period after the introduction of vibroacoustical diagnostics. Similar results were recorded in the Danish refinery department of the Esso Company [12].

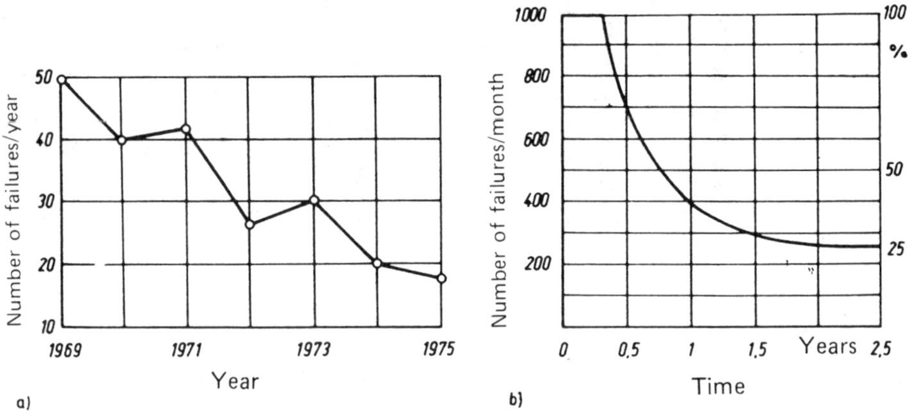

Fig. 1.6 — The decrease of the intensity of failures in a refinery [10] (a), and in one of the industrial plants [11] (b) as the result of the introduction of diagnostics

From the quoted data, one can find reliability characteristics of the group of machines under investigation, such as reliability $R(\theta)$, intensity of failures $\lambda(\theta)$ and the probability of Good Condition $P_Z(\theta)$ (or availability). An estimation of these values where N is the number of machines in operation, and the number of failures is $n(\theta)$ (θ = utilization time) can be found from the following relations [13, 14]:

$$R(\theta) = \frac{N - n(\theta)}{N}, \quad \lambda(\theta) = \frac{n(\theta+\Delta\theta) - n(\theta)}{[N - n(\theta)]\,\Delta\theta} = \frac{\Delta R}{R\,\Delta\theta}, \quad P_z(\theta) = \frac{N - n(\theta)}{N} \tag{1.1}$$

As can be seen from the above relationships, for a group of machines working under continuous operation with regeneration, the determination of reliability $R(\theta)$ is convergent with the estimation of the probability of Good Condition of the object under continuous operation, which can also be identified with the availability factor. Since the original definition of reliability does not include repair, we shall only use the intensity of failures $\lambda(\theta)$ and the probability of Good Condition $P_Z(\theta)$. Note that with a great number of machines (N = great), and a small number of failures (n = small), we can approximately compare the number of failures in time $n(\theta)/\Delta\theta$ to the intensity of failures:

$$\lambda(\theta) \approx \frac{n(\theta)}{\Delta\theta} \quad \left(\text{when } \frac{n(\theta)}{N} \ll 1\right) \tag{1.2}$$

Thus, we can state on the basis of our examples (Fig. 1.6) that the *intensity of failures of a group of machines decreases* from 3 to 4 times after the introduction of vibroacoustical diagnostics. In the latter example (Fig. 1.6(b)) we can calculate even the probability of good condition which for time $\theta = 0$ was

$$P_z(0) = \frac{3000 - 1000}{3000} = 0.66$$

and after two years $P_z(2.5) = 0.91$. So, the introduction of vibroacoustical diagnostics gave over *20% increase of availability* in the utilized group of machines.

Similar data are given by Brüel–Kjaer Company in their diagnostic report [13] quoting an example of a Norwegian chemical plant. In the system of planned preventive maintenance, 200 pumps were overhauled every year, 30 of which were subject to breakdown immediately after overhaul due to errors in repair, and 17 broke down for other reasons during the year. After the introduction of vibroacoustics, the annual number of overhauls decreased to 14, and the time and range of overhauls was known in advance. Calculating on this basis, the probability of good condition without including the already unnecessary planned overhauls is without diagnostics, $P_{ZO} = 0.765$, and after the introduction of diagnostics $P_{ZD} = 0.93$, i.e. an increase of almost 20% in the level of availability in the utilized group of machines. The results then are significant, and worth implementing.

In motor car transport the application of diagnostics also brings measurable results. According to some Soviet investigations [15], as a result of the introduction of diagnostics, one can obtain the following.

(1) Increased car utilization by 15–20%.
(2) Decreased use of fuel by 20%.
(3) Decreased wear and tear up to 20%.
(4) Reduced costs of servicing and current repairs by 15%.
(5) Decreased downtime of vehicles by 30%.

It is obvious that the measurable economic and reliability effects do not appear inexpensive, and according to the authors of the quoted report [9], form the savings, (0.5 + 3% of added value output) 16% must be subtracted for the keeping of diagnostics staff and their measuring equipment. It means that the outlay for vibroacoustical diagnostics is returned by over 50%. The number of the diagnostics staff depends on the needs, and according to the discussed report of Brüel–Kjaer Company [13], the problem is solved in the following ways.

(1) Measurement and analysis of results up to 50 points per month ⇒ one diagnostics engineer.
(2) Measurement and analysis of 50–100 points per month ⇒ one technician + part-time work of one diagnostics engineer for result evaluation only.
(3) Measurement and analysis of 1500–3000 points per month ⇒ 2 technicians + 1 diagnostics engineer for the evaluation of results.

All these vibration diagnostic investigations can be carried out generally with a very simple set of measuring equipment comprised the following.

(1) Vibration meter with the possibility of measuring acceleration, velocity and

displacement with signal output to a tape recorder.

(2) A portable instrument-quality tape-recorder with frequency range at least 10Hz to 2KHz.

(3) A portable spectrum analyser with an X–Y recorder for field and laboratory analyses.

In addition to this set of equipment and staff, knowledge is, of course, necessary and its economic value is proportional to the degree of its range and complexity. The basic knowledge necessary for a diagnostics engineer will be discussed in the following chapters. The different parts of this book can be used by various specialists involved in the evaluation of the condition or the quality of machines, or other technical installation.

2

Generation of vibroacoustical phenomena and wear precesses

2.1 INTRODUCTION

The preceding chapter explained the reasons for the development of vibroacoustical diagnostics and its connections with other methods of machine testing and the aspects of reliability and economy relating to the application of vibroacoustical diagnostics in industry. Having ascertained the use of this science, let us explain the presentation of facts and relations which constitute the vibroacoustical diagnostics. Primarily we have to become acquainted with the basic types of vibroacoustical processes generated during the functioning of machines, the basic processes of machine wear and the mutual relations connecting the two groups of processes accompanying the motion of machines. These interrelations are the basis on which the recognized vibroacoustical methods of the evaluation of the quality and technical condition of machines have been developed.

2.2 SOURCES OF VIBROACOUSTICAL PROCESSES IN MACHINERY

From the point of view of the generation of vibrations and noise as well as other vibroacoustical processes, machines must be regarded as autogenerating systems. This follows from Fig. 1.1, where in response to energy input, we obtain from the output a family of residual processes including vibroacoustical processes as well. Here, the energy supplied to the input of the machine is transformed into a basic technologically useful motion. In general, it can be rotary motion (e.g. electric motors, pumps, etc.), reciprocating motion (piston engines, presses, etc.), or a combination of both basic motions. As can be seen from the examples in parentheses, both basic motions and their syntheses occur in the driving subassemblies of machines as well as in their working parts, which include a whole range of accompanying motions unnecessary from the point of view of the objectives realized by a machine. These accompanying motions (i.e. the dynamic phenomena) are vibration, noise, pulsation of medium and acoustical emission—previously mentioned as the vibroacoustical processes. However, it must be noted here that the generation of vibrations, noise and pulsation does not always evidence a faulty functioning (bad technical condition) of the machine. For example, a piston compressor cannot be perfectly balanced, so some level of vibration must be accepted as the rated one following from the constructional realization of the functional purpose of the machine. Similarly, the dosaged method of medium pressing in the same compressor gives a rated level of medium pulsation, etc. Thus, it is clear that in the majority of cases in machine operation not only the generation of vibroacoustical processes in general is diagnostically significant, but also important are the levels of deviations and the character of the generation of these phenomena in comparison with the standard of a technically efficient machine (it refers also to machines with a resonance character of work). Such an approach implies also the tolerances used in the production of elements, the subassemblies and the whole machines, e.g. unbalance. Having this in mind, let us proceed to the specification and description of the basic types of vibroacoustical processes accompanying the operation of machines.

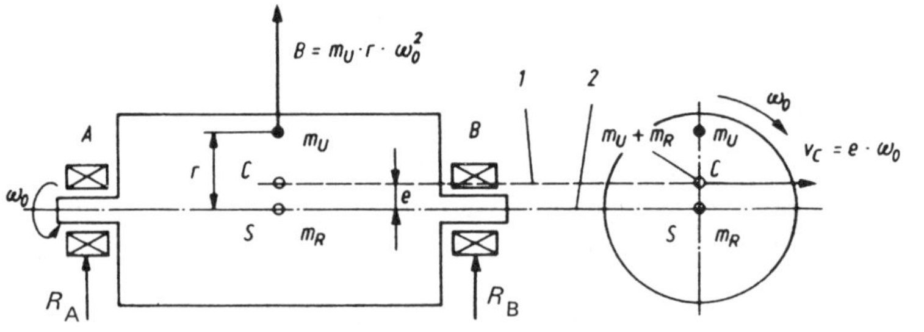

Fig. 2.1 — Rotor statically unbalanced, its mass geometry and reactions

2.2.1 Unbalance

Unbalance is a common feature of all rotational elements, both of the shafts of turbomachines and the crankshafts of piston machines. However, in the first case the unbalance is a side-effect caused by the imperfection of material (e.g. heterogeneity) and the manufacturing process, as well as the effect of wear by use (e.g. erosion).

In the second case, the unbalance of a crankshaft is a characteristic of the machine consciously chosen by the manufacturer; it remains unchanged in use [16] and will not be discussed here. Generally, each unbalance of mass in rotary motion is the source of rotational forces and/or moments of inertial forces which, being transferred through the bearings to the body of the machine, evoke vibrations in the whole system. Measurements of the amplitude and the phase of these vibrations permit us to infer about the type of unbalance as will be discussed below.

Static unbalance

Let us consider the body of a machine rotor on a rigid shaft set on bearings and rotating with an angular velocity of ω_0. Let the mass of the rotor and the shaft be m_W , while on the radius r from the rotation axis in the plane of gravity centre S there is an additional unbalanced mass m_n as shown in Fig. 2.1.

From the figure and on the basis of elementary knowledge of mechanics (see, for example, Leyko [17]) we can easily find the moment of inertia of the rotating rotor by two methods [18]. We have to consider only the unbalanced mass as: $B = m_n r \omega_0^2$. By comparing these two relations we obtain the *proper unbalance*

$$e = \frac{m_n r}{m_n + m_w} \cong \frac{m_n r}{m_w}, \quad \text{since } m_w \gg m_n \tag{2.1}$$

measured in length units, e.g. in millimetres or micrometres.

The value e can be also determined when we know the velocity of the centre of mass C of the unbalanced shaft, since $v_C = e\omega_0$. On the limitation of this velocity the standards are based which determine the permissible unbalance e_d for the particular machine groups (see appendices). The rotational or reaction force of inertia on account of unbalance is taken over by the bearings, since $B = R_A + R_B$. Since the rigidity of the bearings in the machine body and the body itself is always finite, the reactions R_A and R_B will transmit vibrations to the adjacent elements and subassemblies with the rotary frequency $f_0 = \omega_0/2\pi = n/60$ Hz. Thus, by measuring the amplitude of vibrations on the bearings we can estimate indirectly the magnitude of unbalance. Let us note, that with the type of unbalance shown in Fig. 2.1, the vibrations of bearings A and B (bearing supports) will be co-phase (Figs 2.1 and 2.2) which is an essential item permitting us to distinguish this type of unbalance from other faults.

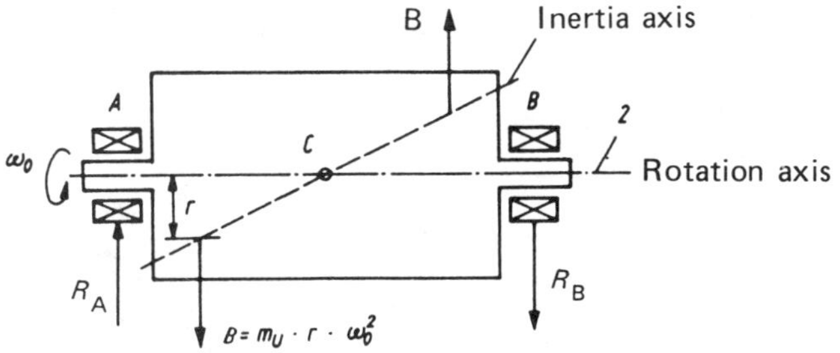

Fig. 2.2 — Moment unbalance of rotor (rotating pair of forces B, B) gives reactions of bearings and their vibrations in counterphase

Moment unbalance

Moment unbalance when the mass allowances are symmetrically and alternatively distributed in relation to the mass centre S (Fig. 2.2). Then the main axis of inertia crosses the axis of rotation in the centre of gravity, and the reactions of bearings are equal and have an opposite sign. In other words, the rotational pair of forces gives bearing reactions which are in the counterphase ($\psi = 180°$). Thus, the vibrations measured on the bearing housings differing in phase by 180° permit us to distinguish easily the *moment* unbalance from the static one.

Quasistatic unbalance

Quasistatic unbalance is created when the unbalanced mass lies beyond the plane of gravity centre. Then the main axis of rotor inertia crosses the axis of rotation beyond the gravity centre S giving proper unbalance e and unequal amplitudinal but cophasic reactions (vibrations) of bearings A and B (Fig. 2.3(a)).

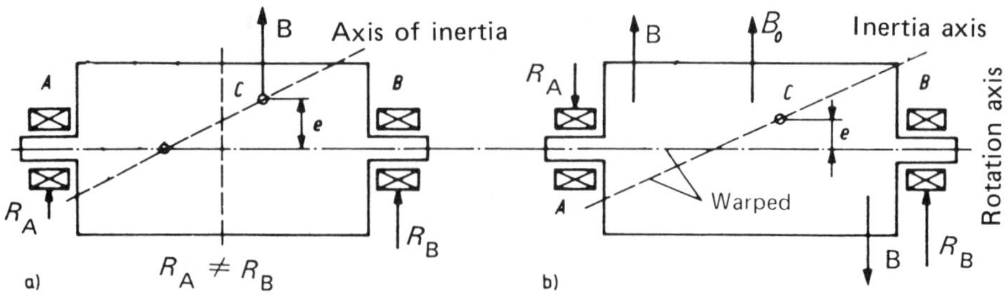

Fig. 2.3 — Quasistatic unbalance (a) and dynamic unbalance (b) of rigid rotors and their reactions (vibrations of bearings)

Dynamic unbalance

According to Łaczkowski [18], dynamic unbalance is a most general unbalanced condition of rotor mass distribution where the main axis of inertia is off centre in relation to the axis of rotation in distance *e* (Fig. 2.3(b)).

From the point of view of mechanics, the arrangement of inertial forces in this case can be reduced to the resultant rotational force vector applied in the gravity centre and the rotational pair of forces equal to the sum of inertia forces moments of unbalance. In such case the reactions (vibrations) of bearings A and B will be alternately different regarding their amplitudes and displaced in the phase by 180°.

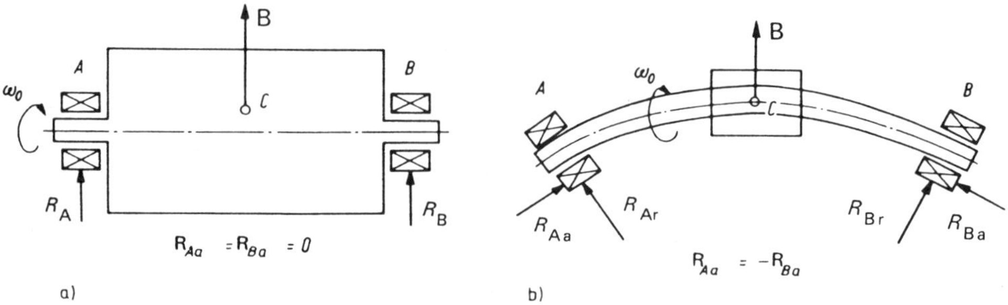

Fig. 2.4 — Differences between unbalance (a) and a bent shaft (b) in the form of counterphase reactions of axial vibrations

Concluding this discussion on unbalance, it is worth reminding the reader that it is a dynamic phenomenon revealed in motion. The names introduced give a better distinction between the particular cases and the recognition of the types of reactions and vibrations of bearing supports. The detection of the type of unbalance facilitates the correction of an existing situation by balancing in one plane in case of moment and dynamic unbalance [18]. The correction of balance becomes unusually essential for machines operating at speeds in the range of their own frequency of the rotor-shaft system vibrations on elastic supports. Such frequency is called the critical frequency $\omega_{kr} = 2\pi f_{kr} = \sqrt{k_z/m_u}$, where k_z is the reduced stiffness of the system, m_z is its mass and, if $\omega_0 \sim \omega_{kr}$, then the amplitude of deflections and vibrations of the shaft can approach large values leading to seizure, failure, etc.

Bent shaft

When a shaft is bent, owing to seizure or heat deformation [19], it produces static or quasistatic unbalance depending on the localization of the flexure centre [20]. Reactions and vibrations normal in bearings are the same here as in Fig. 2.1 or Fig 2.3(a). However, in effect of the spinning of the bent shaft there appear reactions tangential (axial) to the bearings which are mutually opposite regarding their directions (Fig. 2.4). Thus, the measurement of the phase of axial vibrations of bearings and their difference by 180° ensure the detection of the shaft flexure and the correction of the existing situation.

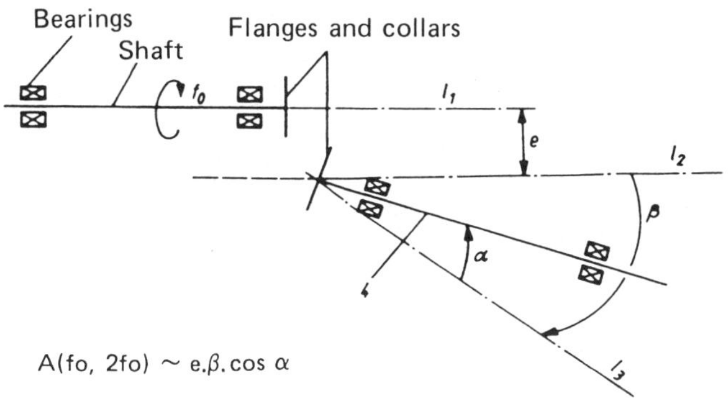

Fig. 2.5 — Possible types of misalignment outlined together and their dimensions, where *e* is misalignment (displacement) radiation, β is angular misalignment in the plane of axis l_1, l_2 and α is oblique to plane l_1, l_2, l_3

2.2.2 Misalignment of shafts, couplings and clearances

The power unit of a machine and its working part form in general separate components joined by the shaft ends. However, this joint is often not aligned and generally one can distinguish three types of misalignment of shafts—radial, angular and oblique (Fig. 2.5)—developed in consequence of manufacturing errors and assembly faults. As indicated by the detailed dynamic analysis [21], the misalignment of shafts is a source of double rotary frequency $2f_0$, ($f_0 = \omega_0/2\pi$). The amplitude of these vibrations is directly proportionally to the radial displacement *e*, and the angular displacement β, and varics with the obliquity parameter α. In radial misalignment the vibrations have a radial character perpendicular to the shafts, and their phases measured on the neighbouring resistance at the coupling are opposite (π = 180°). On the other hand, the angular and oblique misalignments give strong axial components also displaced in phase by π, and furthermore, the amplitude of axial component can exceed the radial one by 1.5 times [20].

Couplings

Couplings joining the shafts of a power unit and the working mechanism have a great influence on the dynamic condition of the machine assembly. The most frequently used ones are the rigid, flexible and gear couplings. The two latter, next to transferring mechanical energy, compensate the assembly's inaccuracies and dynamically separate (vibrations, shocks) both parts of the assembly. *Rigid couplings* consisting of two rigid collars joined by bolts, when correctly aligned, does not give any additional vibrations. *Flexible couplings* by rubber elements, spring elements, etc. is not a source of vibration but, when the assembly is misaligned, we observe vibrations as described above—generally the lesser the higher the flexibility of the coupling [21]. *Gear couplings* permits a greater displacement of shafts than the above axial or angular ones [19]. However, in the majority of cases, when unequally loaded, the coupling itself is the source of vibrations with rotary frequency f_0 and harmonic frequency $z \cdot f_0$ according to the number of teeth of the coupling z.

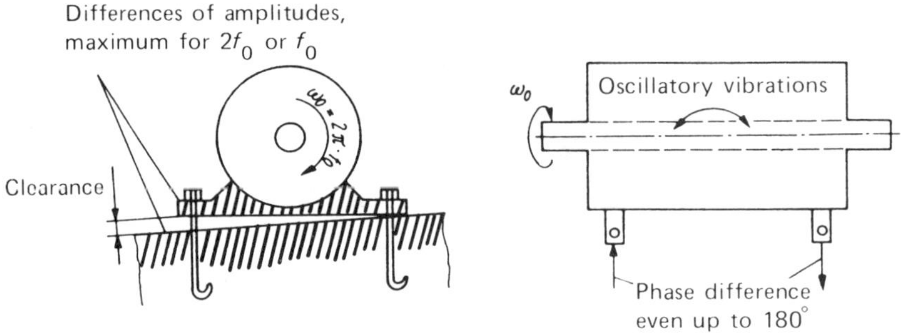

Fig. 2.6 — Assembly clearances in the vertical plane (a) and horizontal plane (b), and methods of their detection

Clearances in the mounting

In clearances in the mounting of the machine assembly, the misalignment of the driving shafts and the working mechanism line forces a spatial motion of both members of the assembly with the highest amplitudes in the horizontal plane. This type of vibration is often caused by loosened bolts anchoring the units of the assembly to the foundation. Such loosening, depends on the vertical or horizontal direction and gives vibrations with impacts against the limiters (consisting of the head or mandrel of the anchoring bolt) which leads to the truncation of the bolt and to a breakdown of the assembly. The detection of a clearance in the early phase of development is possible by the measurement of the differences between the amplitudes and the vibration phase, the anchoring bolt and the machine lug, or the phase between two assembly lugs (Fig. 2.6). Depending on the number of impacts for a given period, the strongest vibrations caused by clearance can develop for the rotary frequency f_0 or the double frequency $2f_0$ [22].

As indicated in the diagnostic guide [19], horizontal loosening of assembly bolts gives differences in the phase of the horizontal vibrations of machine lugs. At the same time, when the loosening is on one side, it is best to observe the vibrations for frequency $2f_0$, and when the loosening is on both sides, the frequency $1f_0$ should be watched. The same refers to the loosening on both sides in the perpendicular direction.

2.2.3 Critical speed of the second order

The unbalance discussed in Section 2.1.1 can be the cause of critical vibrations of the first order [18] since, as has been mentioned, it will take place if $\omega_0 \sim \omega_{kr} = \sqrt{(k_z/m_z)}$. However, the same resonance frequency of the system ω_{kr} can be induced in certain cases when the relation $2\omega_0 \sim \omega_{kr}$ occurs. Then $\omega_0 = \omega_{kr}/2$ is called the *critical* frequency *of the second order* of the rotor shaft system on elastic supports. As explained by den Hartog [23], it occurs due to the parametric vibrations in the system where the elastic parameter k is the function of the rotation angle $\psi = \omega_0 t$. For the shaft rotor unit, such situations can occur in two cases: (a) in the asymmetry of the shaft cross-section in the presence of keys, flats, grooves for the winding, etc.; (b) when the moment of inertia of the shaft cross-section creating flexural rigidity EI_z = Young's modulus ×-moment of inertia in respect of the neutral axis) takes the maximal value twice during one revolution of the shaft (Fig. 2.7). This periodic change of stiffness supplies energy for the vibrations with the frequency $\omega_{kr} / 2$ causing dangerous resonance vibrations.

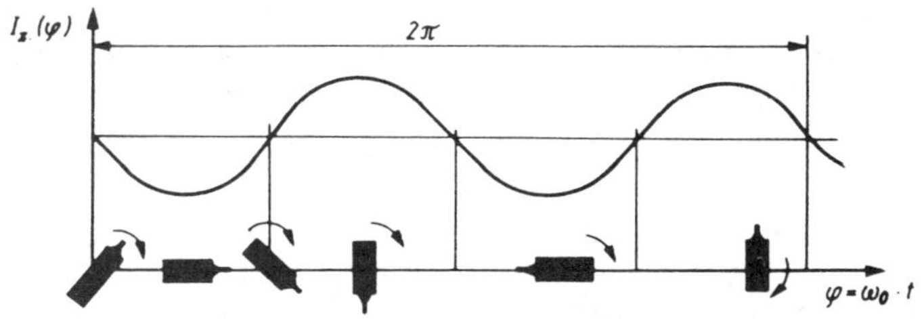

Fig. 2.7 — Change of the inertia moment of shaft cross-section as the cause of parametric vibrations and critical rotations of the second order [23]

The second possibility of parametric vibrations is connected with the unbalance of the rotor rotating around the horizontal axis. The gravity force of the rotor G on the unbalance arm e during one revolution $\psi = 2\pi$ accelerates the motion on one occasion and on another retards it, effecting an elastic contact force which depends on the position of $k = k(\psi) = k(2\omega_0 t)$. When the rotations of the rotor approach $\omega_0 = \omega_{kr}/2$, then a parametric resonance occurs in the system and the amplitudes of the rotor vibrations can reach very high values.

As indicated by a detailed analysis [23], a significantly more probable one is the case of critical vibrations of the second order caused by the asymmetry of the cross-section, and in doubtful cases the balancing of the rotor will resolve the question of the vibration cause.

2.2.4 Self-excited vibrations of bearings

A frequent phenomenom in machines set on slide bearings is self-excited subharmonic vibrations of the order of $(0.1–1) f_0$. These are more dangerous than unbalance. Since the shaft is subject to vibration, it means that specific cross-sections of the shaft are alternately compressed and stretched (flexural vibrations), while in the case of unbalance we have to do with dynamic spinning of a bent shaft line. Such vibrations are caused by unstable conditions of the bearings depending on clearance, pressure of lubricant, viscosity of oil, speed of rotation, and loading. Generally, the greater the journal-bushing relation in the bearing, the higher the possibility of initiating self-excited vibrations. Instability can be created here which is called the *oil whirl* [24, 25] or *hydrodynamic precession*.

The phenomenon of oil whirl depends on a destabilizing influence of unsymmetric distribution of pressure in the bearing in which the centre of the journal was subject to displacement in relation to the centre of the bearing bushing. The cause of this displacement can be a large clearance, unbalance, loading conditions, etc. Self-excited vibration created in this way with a frequency of about $0.5 f_0$ can very easily develop into dangerous amplitudes if this frequency is close to the resonance frequency of the system consisting of shaft-, rotor- and bearing-mounted supports. This phenomenon is known as *resonance whirl* or *resonance precession*. In both cases, the motion trajectory of the journal centre around the centre of the bearing bushing has a direction agreeing with the direction of rotation f_0. The opposite direction of the self-excited motion occurs in *frictional whirl*, being the effect of a high clearance, insufficient lubrication and momentary seizure of the journal in the bearing bushing. In such a case, the self-excited vibrations of submultiple and supermultiple $1/2\, f_0$ continue to develop. These cases of self-excited vibrations of turbomachines are the result of instability of the journal bushing system of the slide bearing.

In these machines, vibrations due to instability can be induced by the material of the shaft (called *hysteresis whirl*) [23, Chapter 7] with the participation of the turbomachine rotor seals [25, 19, Chapter 2] and the influence of a flexural vibrating rotor on the shaft in an axial flux of working agent at a high velocity [25]. In all these cases, the frequency of the induced vibrations is subharmonic in relation to the rotations within $(0.1–1) \times f_0$, are rarely superharmonic in the range of $1.5\, f_0$.

2.2.5 Rolling element bearings as a source of vibration

The bearing is the most common machine element and its main task is to transfer safely the working load from the rotating element to the body with small motion resistance and good positioning of the rotating working element. On account of the kinematics of the element we classify bearings into sliding and rolling, while on account of the method of transferring load from the rotor to the body of the bearing, we classify them into radial (journal), axial (thrust), and angular (oblique). Bearings can be modelled with substitute mass, stiffness and damping. However, such an approach is justified only in cases of the previously discussed slide bearing being beyond the sphere of self-excited vibrations. The most important sources of vibration generation in rolling bearings are *construction, manufacturing and utilization.*

The constructional sources of vibration follow from the vital work performed by a rolling bearing (i.e. the periodically variable number of rolling elements in the bearing contact transferring the load) and from the transfer of load with the help of a movable element in point contact (balls) or linear contact (rollers). Owing to the first parametric vibrations, caused by the periodic change of the number of elements in the bearing contact causing a change in the rigidity of the bearing similar to the asymmetry of the shaft cross-section discussed in Section 2.1.3.

The second feature—the moveable load—is the source of waves running forwards and backwards from the contact area, and with radial speeds of the rolling element bearing comparable to the speed of sound in the material of the raceway it may produce significant self-excited vibrations.

Manufacturing causes are related to the presence of deviations in form and dimension and with the manufacturing and assembly faults. The range of possible deviations and faults varies from roughness of the bearing element through waviness, point faults to run-out and clearance in the bearing. The causes of these vibrations are summed up in Fig. 2.8 and more detailed information can be found elsewhere [26].

Finally, use causes vibrations to result from the wear and friction processes and fatigue, generally increasing the existing deviations of form and dimension and creating new point faults and surface faults (pitting) on the raceways and the rolling elements of the bearing. Hence, the cumulative data presented in Fig. 2.8 are also useful for the Fig.

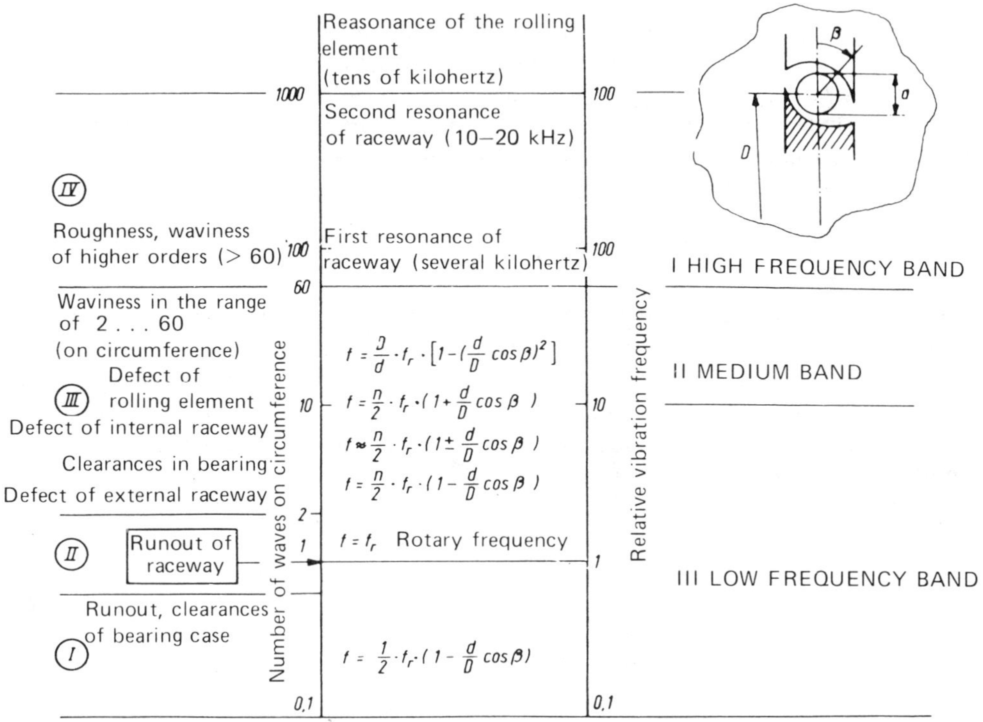

2.8 — Schematic prepresentation of the main reasons of rolling element bearing vibrations in the scale of frequency in relation to f_r with an indication of the corresponding formulae and vibration ranges [7]

presentation of causes of rolling element bearing vibrations.

2.2.6 Toothed gears

In machine construction there is frequently a need to change the rotational speed or the driving torque between the motor and the working machine. In the majority of cases toothed gears are used, particularly when it is necessary to transfer high power. The gears, due to discrete transfer of load by the successive meshing teeth, belong to the essential sources of vibrations. Here, the periodically variable number of teeth in the unit causes periodical fluctuations of meshing rigidity which, as already explained for the shaft with variable rigidity and for rolling element bearings, is the cause of the development of parametric vibrations with characteristic modulating effects [6, Chapter 6]. The structure of gear vibrations is very complex, particularly because it includes vibrational effects caused by manufacturing errors and assembly faults. To illustrate this fact, let us consider the simplest one-step gear with parallel axis and outline the form of vibration spectrum which can be recorded on the gear housing. In the simplest case, the mechanical elements are : slow-speed input shaft with frequency f_{01}, two toothed wheels with teeth number z_1 and z_2 and high- speed output shaft f_{02}. Each of the wheels can be set more or less eccentrically and with some misalignment.

In the low-frequency part of the spectrum, there are successive components f_{01}, f_{02} as the effect of the eccentric and several harmonics of these frequencies as the result of misalignment. The successive components of vibrations will be the harmonic frequencies of meshing $f_z = z_1 f_{01} = z_2 f_{02}$, the significance of which can reach $5f_z$ in the gear spectrum. On the other hand, the eccentricity of the wheels causes a periodical change of the kinematics of meshing, giving in turn the effect of modulation and side bands around each harmonic f_z which can be obtained by:

$$f_z = nf_z \pm l\, f_{0_1} \pm m f_{0_2}\,, \qquad n,\, l,\, m = 1, 2, 3, \ldots \tag{2.2}$$

The vibration spectrum of a hypothetical gear housing has been presented in Fig. 2.9, and more detailed data about the causes and the character of the vibrations can be found, for example, Laczkowski [19] and the author [67]. Here, it is worth stressing that the utilizational wear of the gear—point wear (e.g. breaking teeth) and superficial wear (e.g. pitting on the working surface of teeth)—will increase the amplitude and complicate the spectral characteristics of gear vibrations.

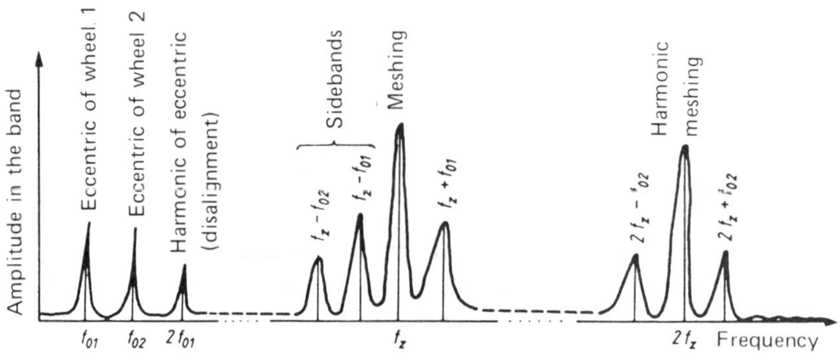

Fig. 2.9 — Hypothetical spectrum of vibrations in one-step gear with typical manufacturing and assembly defects

2.2.7 Magnetic sources of machine vibrations

One of the important sources of power is the electric current transformed by the electric motors into motion. The essence of the transformation consists here in the utilization of the forces in the gap between two magneto-electric circuits. Let the angular position of the motor be denoted by ψ, and let the air gap between the rotor and the stator be $d(\psi, t)$. Let $B(\psi, t)$ be the magnetic induction in the gap which, as we know [27] is the product of the magnetomotoric force $\Phi(\psi, t)$ and the conductivity of gap $\Lambda(\psi, t)$ which in turn is inversely proportional to the thickness of the gap $\Lambda(\psi, t) \sim d^{-1}(\psi, t)$. Then the normal radial force $p(\psi, t)$ in the gap can be expressed in the form:

$$p(\psi, t) = \frac{B^2(\psi, t)}{2\mu_0} = \frac{1}{2\mu_0}\,\Phi^2(\psi, t)\,\Lambda^2(\psi, t) \tag{2.3}$$

where μ_0 is the magnetic conductivity of air in the gap. The magnetomotoric force changes with the frequency of network supply ω_z ; hence $\Phi(\psi, t) = b(\psi) \cos\omega_z t$ while the conductivity of the gap Λ depends on the number of poles p of the stator and the rotor, and in the first approximation it can be approximated by the expression: $\Lambda(\psi, t) \cong \lambda_0 +$

$\lambda_1 \cos\omega_0 pt$, where ω_0 is the rotary frequency of the motor and p is the number of pole pairs. Substituting Equation (2.3), we find:

$$p(\psi, t) = \frac{b^2(\psi)}{2\mu_0} \frac{1+\cos(2\omega_z t)}{2}$$
$$\left(\lambda_0 + 2\lambda_0 \lambda_1 \cos\omega_0 pt\right) + \frac{\lambda_1^2\left[1+\cos\left(2\omega_0 pt\right)\right]}{2} \tag{2.4}$$

Therefore, the force exciting the vibrations of motor elements will have components with the following frequencies.

(1) Double supply frequency $2\omega_z$ which, for ferromagnetic materials, is usually called *magnetostriction*.
(2) Frequency depending on the rotation ω_0 and the number of pole pairs p or $p\omega_0$, and its second harmonic $2p\omega_0$.
(3) Summing–differential frequencies $2p\omega_z \pm p\omega_0$ and $2\omega_z \pm 2p\omega_0$.

In the simplest case—the spectrum of force $p(\psi, t)$—we will find as many as six spectral components forcing vibrations. When we realize that the supply current is not always a pure sinusoid, and the magnetic conductivity of the gap in general cannot be approximated by one harmonic, then we will find how rich the spectrum of motor vibrations must be. It is rich in a *sequence of supply harmonics* $l\omega_z$ and in a sequence of *summing–differential* harmonics $l\omega_z \pm np\omega_0$ ($l, n = 0, 1, 2, \ldots$) following from the number of poles. If we add to this the conversion of manufacturing errors and assembly and utilization faults (ovality of stator, asymmetry of field, etc.) into harmonic magnetomotoric forces, then we will have the proper spectral image of the sources of electric motor vibrations [6]. Hence, the forcing of such a dense frequency spectrum always meets with resonances of motor construction giving significant vibration amplitudes.

2.2.8 Motion of the working medium

In fluid-flow machines whose function consists in the change of energy of the working agent (e.g. of a compressor, pump, fan, blower) or in the utilization of the energy of the agent (e.g. turbines, pneumohydraulic systems and even internal combustion engines) the general principle consists in the creation or utilization of the *heterogeneous distribution of velocity* and the pressure of the agent. This process of the aimed creation of heterogeneity takes place in the machine with a constant interaction of three basic elements of fluid-flow machines: (a) the body or housing of the working space of the machine; (b) the working agent (fluid, gas); and (c) the movable part in the form of rotor or piston. Thus, each change of the energy of agent, and particularly its pressure and/or velocity, is in the same way received by the body and the movable mechanism of the machine. We know from the mechanics of fluids that if a body moves with a relative velocity $V(t)$, then the pressure acting on it is:

$$p(t) = p_{st} + \rho\frac{V^2(t)}{2} = p_{st} + \rho\frac{\left[V_0 + \bar{V}(t)\right]^2}{2} \tag{2.5}$$

where p_{st} is the static pressure in the stream, $V(t) = V_0 + \bar{V}(t)$ is the momentary value of the velocity to the analysed element of the stream with pulsatory component $V(t)$, ρ is the density of the working agent.

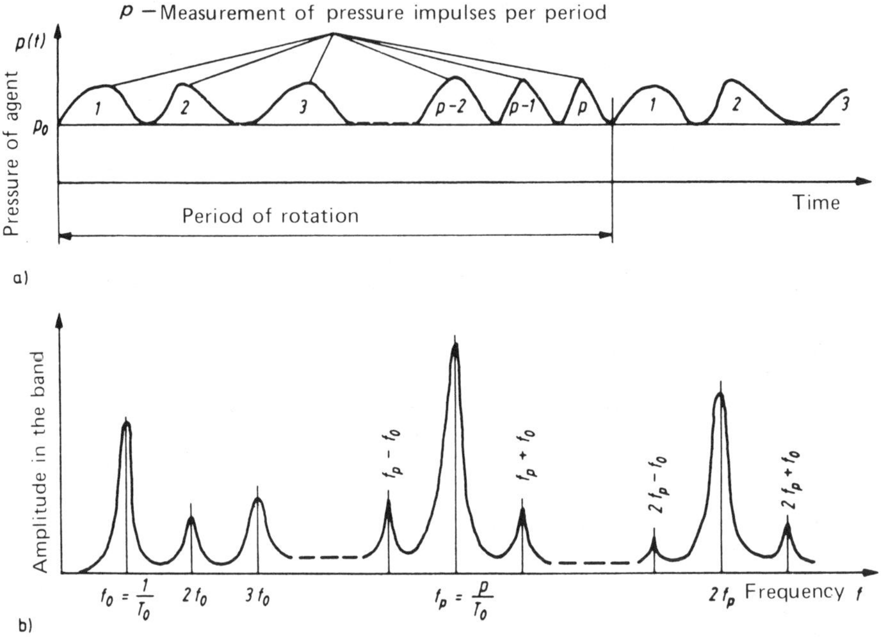

Fig. 2.10 — Sequence of periodical impulses of medium pulsation in fluid-flow machines (a) and its spectral characteristics (b)

Pulsatory component $\overline{V(t)}$ develops in fluid-flow machines as the effect of motion of blades, motion of piston, opening and closing of the working spaces, etc. Depending on machine operation, it can be the only velocity component of the agent (as in piston machines) or can modulate the component of the uniform motion of agent V_0. However, in each case of the constructional way of machine operation, there exists a significant pulsatory component which induces vibration of the movable and immovable elements of fluid-flow machines. Hence, each deviation from the designed course of pulsation $\Delta V(t)$ resulting from manufacturing and assembly deviations (e.g. irregular pitch of rotor blades) or from utilization reasons (e.g. fatigue break, blade erosion, valve wear, etc.) will change the course of forcing and the vibrations of the constant and movable elements of the machine. The spectral structure of forcing and vibrations in fluid-flow machines is very complex and it is best to analyse the forcing as a periodical sequence of impulses following from the number p of blades, etc., repeated with the rotation period of rotor or crankshaft (Fig. 2.10). The neighbouring impulses within the period $T_0 = (2\pi/\omega)_0$ are here similar but not identical. Hence, the basic frequency is $f_0 = \omega_0/2\pi = l\,/\,T_0$ with the successive harmonics until the blade frequency $f_p = pf_0$ and its successive harmonics $f = lpf_0$ ($l = 1, 3, \ldots$) as in Fig. 2.10(b). Generally, one can say that the more the impulses 1, ..., p are similar to each other in period T_0, the higher harmonic frequencies f_p dominate in the spectrum. Furthermore, if the sequence of similar impulses comes close to the course of the type sin $p\omega_0$ or cos $p\omega_0 t$, then the higher harmonics f_p disappear and there remains only the frequency connected with the number of blades or cylinders.

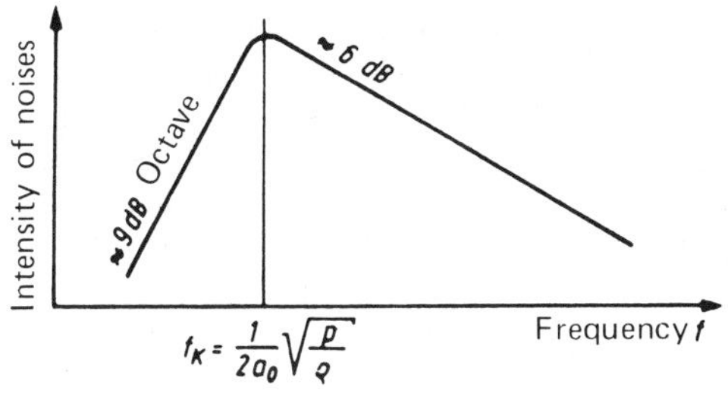

Fig. 2.11 — Frequency spectrum of cavitation noises according to Ross [30]

2.2.9 Cavitation

Cavitation is one of the most powerful sources of vibration and sound in fluid. It is created when the pressure in the stream of fluid P becomes smaller than the pressure of its saturated vapour in the given temperature P_d, i.e. when $P < P_d$. Conditions for such a pressure drop occur frequently: in the installations of pipelines, particularly in nozzles, valves, bends, in flows around immovable obstacles, in conditions of rotational flow and turbulent flow, in fluid flow from off a blade, pump, turbine, power screw, etc., and even in slide bearings [28]. In these cases, there occurs on the working mechanism, or in the installation of a fluid-flow machine, the *creation of bubbles* of saturated vapour followed by rapid increase and disintegration. This causes strong vibrations in the fluid stream and *cavitational erosion* of the machine element, on whose surface the cavitational bubbles disintegrate. It is believed [29] that these disintegrations are rapid enough to be the source of shock waves in the fluid, which by acting on the surface of the machine element lead to a quick cavitational erosion. According to Ross [30], the acoustical power of the cavitation source is proportional to the pressure of disintegration P and the volumetric velocity of the creation of cavitation bubbles. The spectrum of cavitational noise has a wide band character with a maximum frequency equal to:

$$f_k = (1 / 2a_0) \sqrt{(P/\rho)} \tag{2.6}$$

where a_0 id the radius of the bubble in the moment of disintegration, P is the pressure of disintegration, and ρ is the density of fluid.

Depending on the size of bubbles, this spectrum has its maximum between 10–20 kHz with a drop in low frequencies of the order of 9dB per octave and in high frequencies of the order of 6dB per octave (Fig. 2.11). Thus, with the maximum cavitation frequency around 15kHz, the participation of this phenomenon at 30 kHz as well as 5kHz can be rather significant and can mask other vibroacoustical signals and symptoms of machine wear.

2.2.10 Frictional vibrations

These include all vibrational phenomena which occur when two contacting surfaces experience relative motion, in general in the presence of a lubricating agent. So it is a case of cooperation of elements in the majority of kinematic pairs of mechanisms in which *wear by friction* occurs. There are many theories of friction [29], and according to Moore [31] some causes are as follows mechanical joining, molecular attraction, electrostatic attraction and levelling and truncating of microroughness. Let us imagine that the elements of a kinematic pair move with a relative velocity $v = 1$ m/s and have a roughness wavelength of the order $\lambda_a = r_a = 10$ μm. During the motion of this kinematic pair there occurs the truncation of the most 'protruding' microroughnesses, so let us assume that every tenth microroughness is truncated. In this case, the mean frequency of microimpacts of both surfaces against each other during such truncation of microroughnesses will be:

$$f_t \approx \frac{v}{10 r_a} = 10^4 \text{ Hz} = 10 \text{ kHz} \tag{2.7}$$

Since the particular events of micro-abrasion last for a very short time and the process of each truncation is a rapid one, the spectrum of frictional vibrations will be very wide with an amplitude decreasing in the direction of low and high frequencies, with a maximum at frequency f_t as found in the case of cavitation (Fig. 2.11). Therefore, some troubles may sometimes occur with distinguishing the causes of vibrations, particularly when the cavitation is weak.

2.2.11 Acoustic emission

Acoustic emission is an elastic wave due to a violent release of energy accumulated in the material by the propagating microdamage (increase of microgaps, movement of dislocation groups) in the material. The notion of material should be understood here in a comprehensive way starting with the ground, rock mass and materials used in construction to plastic materials and composites of artificial fibres. These materials, have uneven distribution of elastic energy (residual stress) and a certain *level of defects in the structure* and even some micro- and macrodamage. Now, when an external cause appears to change this situation, then in one or in any areas of the material there will occur a release of a portion of elastic energy, and a sequence of decaying elastic waves is emitted. The cause of the change of the state of equilibrium in the material can be the change of stress or temperature. Also, it can be the progressing process of corrosion, and even radiation, i.e. the bombarding by a stream of particles. When we assume that the structure of the surface layer acquired during the manufacturing stage, and the interior of the material in the investigated element, was good (i.e agreeing with the accepted quality standards of manufacturing) then *the appearance of acoustical emission* should be regarded as a signal of property degradation of the given construction element [32].

The acoustical emission can be also seen from the point of view of changes in the material. Among others they include the following.

(1) The movement of vacancies and dislocations, slips on the grain boundaries—in micro-areas with high stress around the plasticity limit of the material.
(2) Combined dislocations, creation of gaps and their development—the latter is a strong source of acoustical emission.
(3) Phasic changes in the crystalline structure of the material.

It follows that acoustical emission can be a good tool for the study of the physical processes of material destruction and, on the other hand, can be used as the tool for the detection of faults and internal defects decreasing the strength of the given construction element. As an illustration of these possibilities, Fig. 2.12 shows a sample of material with an internal defect subjected to an additional load of $\Delta\sigma$.

When the increase of stress by $\Delta\sigma$ is a sufficient cause to disturb the internal energy balance, then concentration of stress occurs around the defect; the stress will exceed the

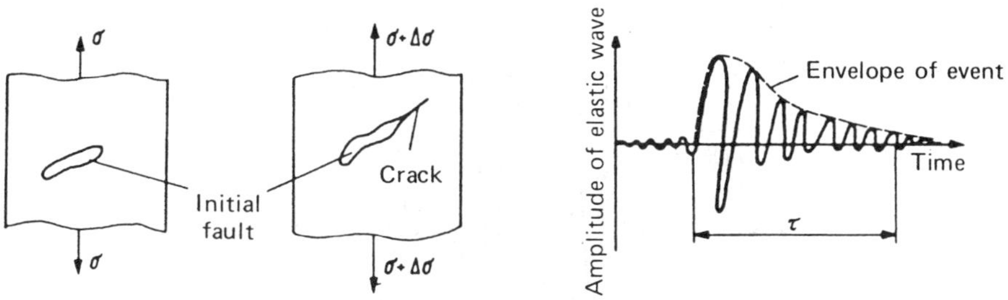

Fig. 2.12 — Schematic presentation of a gap in a sample and the generation of acoustical emission [32]

local plasticity limits, and a propagation of the microgap and emission of a packet of elastic waves or acoustical emission will follow. In general, the amount of elastic energy released in this way in the unit of time is proportional to the volume of the sample, the speed of deformation, and the amount of microdamage created in this way. In the estimated calculations one can assume that the gap propagates with longitudinal wave velocity $c = \sqrt{(E / \rho)}/$ where E is Young's modulus and ρ is the density of material. Thus, the increment of gap length Δl can be connected with the time of cracking τ and the frequency band Δf occupied by acoustical emission impulse. According to Ivanov [33] the following dependence is valid:

$$2\Delta l = c\tau = c / \Delta f \tag{2.8}$$

Oscillation of the acoustical emission impulse of a time length τ depend on the properties of the propagation track and the geometry of the place of impulse generation; they involve not only the reflections from the element limiters but also the inclusions and internal heterogeneities. For example, let us assume that the increment of the gap in the cracking process was Δl = 1 mm and for steel $c = 5 \times 10^3$ m/s = 5×10^6 mm/s. From Equation (2.8) we have $\tau = 0.4 \times 10^{-6}$ seconds, Δf = 250 kHz. In reality, the increments of the gap lengths are considerably smaller and, hence, the spectral range of acoustical emission will be significantly wider, reaching 1MHz or more. For a more detailed estimation of the acoustical emission spectrum it is advisable to consult Tverskoi [34].

In measurements and examinations of acoustical emission rate we use the total number of impulses or events, or the acoustical emission rate which is the number of events in a determined unit of time. Also, in investigations, the acoustical emission energy is used, which is the field under the envelope of the amplitude squared. These values are presented schematically in Fig. 2.13 as an example of acoustical emission impulse.

For a determined method of observation of the acoustical emission phenomenon there exists a strict proportion between the number of summations and the number of events; hence, the relations given below are correct for both cases of the valuation of the acoustical emission phenomenon.

The evaluation of acoustical emission requires the introduction of another factor K, the stress intensity factor which in its threshold value K_{CO}, is a constant characteristic for the given type of material. The formula for the stress intensity factor depends on the type of stress condition and the postulated type of gap. In general, one can assume that its value is proportional to stress σ and the radical from the length of the gap l, $K \sim \sigma\sqrt{l}$ [35].

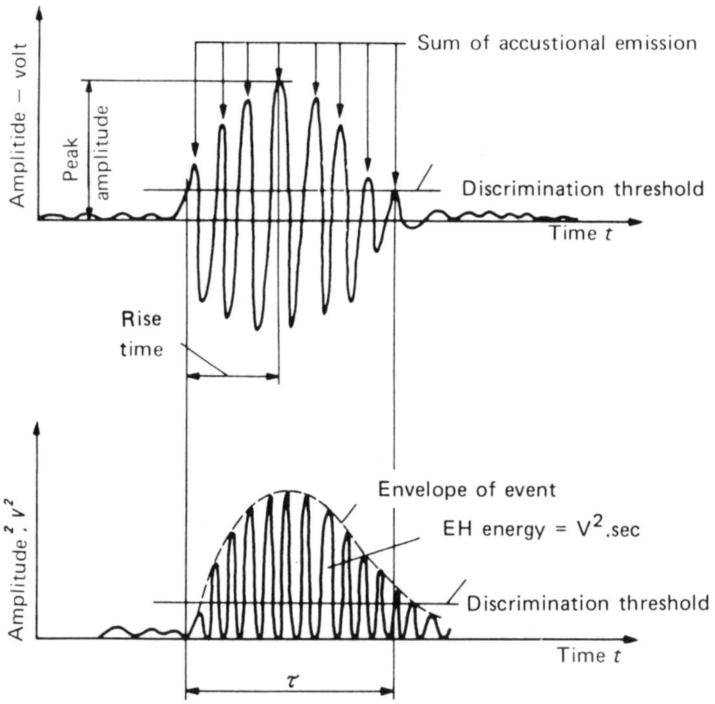

Fig. 2.13 — The most important parameters of the description and valuation of the acoustical emission phenomenon where Σ is the number of sums, event, event energy in V^2 s duration of event, τ [32]

Generally, the greater the value of K, the easier the propagation of existing and initiated microdamage. Hence, the sums of events Σ can be expressed by the following formula given by Dunegan [36]:

$$\Sigma = A\,(K^m - K^m{}_{co}) \tag{2.9}$$

where A is a certain material constant, and m is an exponent depending on the type of material.

Similarly, the rate of emission F recorded with the same process of loading can be written:

$$F = \Delta\Sigma/\Delta t = B\,(K^m - K^m{}_{co}) \tag{2.10}$$

where B is another material constant.

In oscilating loading with a range of stress variability $\Delta\sigma$ we can define the range of stress intensity $\Delta K = K_{max} - K_{min}$. Then the rate of emission per cycle of variable loading N or in the unit of time can be expressed by:

$$F = \Delta\,\Sigma/\Delta t = C_1\,(\Delta K)^{S_1} \text{ or } \Delta\Sigma/Nt = C_2\,(\Delta K)^{S_2} \tag{2.11}$$

with constants C and s depending on the type of material.

From the quoted relations it follows that the acoustical emission phenomenon describes the destruction of material, either the single occasional one or the long-lasting one of the fatigue type. We can postulate that similar relations are true for other types of material wear utilized in the studies on the cohesion and loadbearing structure elements (e.g. bridges and drilling platforms) and machine constructions (e.g. machine bodies and rotors).

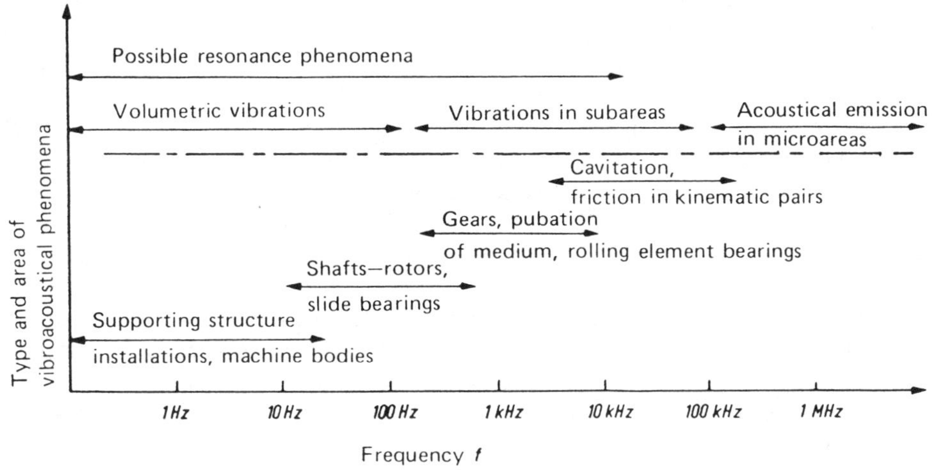

Fig. 2.14 — Synthetic presentation of spectral ranges of vibroacoustical phenomena in machines and installations

2.2.12 The range of vibroacoustical in machinery and structures

Concluding the discussion on the sources and characteristics of vibroacoustical phenomena observed in machines, it is now appropriate to synthesize the acquired information and to realize the global spectral ranges of occurrence. An attempt at such a synthesis in the field of frequency is presented in Fig. 2.14.

As we can see in the figure, we have isolated elements and phenomena occurring in them in the order of decreasing linear magnitudes covering two frequency decades. This classification refers primarily to the extent of vibration generation, in the range of metres from the body and the rotors of machines, and in the range of millimetres and micrometres in frictional vibrations and acoustical emission, implies a conclusion about the great amount of information contained in the vibroacoustical processes, information about the dynamic condition and technical condition of the machine, and the wear processes taking place. These problems will be the subject of consideration in the following sections.

2.3 MACHINE WEAR AND VIBROACOUSTICAL PHENOMENA

2.3.1 Introduction

The consideration about the generation of vibroacoustical phenomena and the place of diagnostics in machine life carried out so far do not raise any doubts as far as the justification of the phenomena of the constructional and the checkout diagnostics are concerned. However, *the wear process* of machines and installations and its connection with the vibration and general vibroacoustical processes require clarification. By wear we understand, according to Będkowski [8] the loss of functional properties of the object. *This property degradation* (gradual or sudden) can be recovered by a regeneration of the object comprising regulation, repair of elements replacement of elements, supply of fuel, lubrication and the removal of wear products.

The wear process is not a homogeneously structured phenomenon, and without going into details (see, for example, Łuczak and Mazur [29]) for our purpose of vibroacoustical diagnostics we can classify it into four degradation processes which differ physically. First, we have *fatigue wear* leading to the *loss of integrity* of the construction elements and occurring on account of large dynamic and oscillation stresses, i.e. vibratory stresses. The second successive process is *friction wear* in all its forms taking place in kinematic pairs and leading to *mass decrements–clearances*. The rheological properties of construction materials are the cause of the third way of *wear by creep* leading to changes in the dimensions and forms of the stressed construction elements. Also important as a fourth consideration is the *erosive wear* of machine and construction elements consisting of physical and chemical degradation and mass decrement of material. Let us discuss in detail the specific ways of machine wear taking into consideration the cause-and-effect connection with vibration and other vibroacoustical processes.

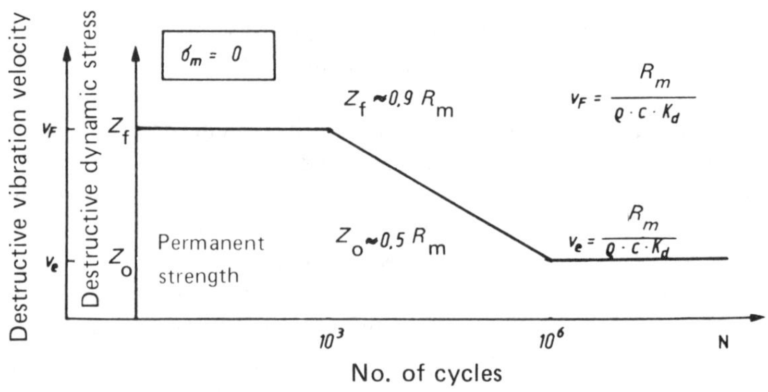

Fig. 2.15 — Generalized Wöhler's diagram for steel alloys in categories of dynamic destructive stress and the corresponding nondestructive vibration velocity for $\sigma_m = 0$ [39]

2.3.2 Fatigue wear

The measure of mechanical loading of the construction element is the stress level in it, which almost always can be divided into two components: the mean working stress σ_m, and $\sigma_a(t)$ the dynamic vibrational stress with the mean value equal to zero. Thus, the total stress will be:

$$\sigma_c(t) = \sigma_m + \sigma_a(t) \tag{2.12}$$

From strength of materials literature we know that when the dynamic stress is equal to zero, $\sigma_a(t) = 0$, and the working stress σ_m does not exceed the ultimate strength $\sigma_c = \sigma_m < R_m$, then the life (*ability to transfer the load in a present time and conditions*) of the element is unlimited. The cause of the limited life is the occurrence of dynamic stress with an amplitude higher than the fatigue limit Z_{g0}, ($\sigma_a > Z_{g0}$). This situation is well illustrated by Wöhler's diagram [37, 38] which for ferrous alloys has the form as in Fig. 2.15.

It is seen that the dynamic stress in the element vibrating in a stationary regime can be expressed in terms of the vibration velocity of the element [39]:

$$\sigma_a = \hat{v}\, \rho\, cK_d \tag{2.13}$$

where v is the peak amplitude of vibration velocity measured in the place of the maximum dynamic deformations, ρ is the density of the mass of material, $c = \sqrt{(E/\rho)}$ is its sound velocity, and K_d is the dynamic factor depending on the concentration of dynamic energy (K_d = 1–3 generally, $K_d < 1$ for small elements, $K_d > 1$ for elements with a great span in relation to the length of the vibration wave in the material). Using the above dependence we can convert the fatigue limit stress Z_{g_0} into the velocity limit of vibrations v_e of limiting stress:

$$v_e = \frac{Z_{g_0}}{\rho\, cK_d}, \qquad \sigma_m \equiv 0 \tag{2.14}$$

and at the same time this form refers to material samples. For construction elements the value given in Equation (2.14) must be divided by an adequate factor of safety, but we will not discuss it further. Generally, the working stress of machine elements differs from zero, $\sigma_m \neq 0$; hence, this fact must be taken into consideration with the help of the simplest hypothesis, e.g. the Goadman's hypothesis [38], where we obtain:

$$v_{em} = \left(1 - \frac{\sigma_m}{R_m}\right) v_e = \left(1 - \frac{\sigma_m}{R_m}\right) \frac{Z_{g_0}}{\rho\, cK_d} \tag{2.15}$$

where R_m is the ultimate strength of material.

From the above-quoted experimental facts and their simplified analytical notation there follows distinctly a causative connection between the phenomenon of fatigue and vibrations. Vibrations are the cause of the occurrence of the phenomenon, and at the same time they are a measure of its intensity (see, for example, Equation (2.1.5)), since the higher the amplitude of vibrations, the smaller the life of the vibrating element. With each cycle of variable load there follows a certain increment dl of the length of the fatigue crack which can be expressed [35]:

$$\frac{dl}{dN} = C_3(\Delta K)^g \cong C_3(\Delta \sigma \sqrt{l})^g \tag{2.16}$$

where C_3 and g are material constants and ΔK is the known range of stress intensity $\Delta k = \Delta\sigma\sqrt{l}$. According to Kocaňda [40], equation (2.16) with a constant σ_a can have a slightly different form:

$$\frac{dl}{dN} = C_3\, \sigma_{\max}^n = C_3(\sigma_m + \sigma_a)^n$$

where σ_m is the working stress and σ_a is the amplitude of dynamic stress.

The cumulative effect of crack increase is the function of the number of cycles N (or wear time $\Theta = N{\cdot}f$, where f is the frequency of the loading) is presented in Fig. 2.16 [41].

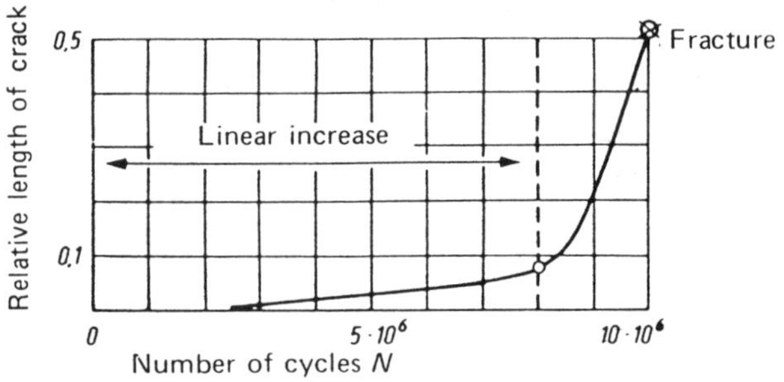

Fig. 2.16 — Depth of fatigue gap in the material, in the function of the number of stress changes or the time of utilization [41]

It can be seen that the loss of integrity by the element takes place almost immediately if the length of the fatigue gap reaches one-half of the critical dimension of the element. In the process of cracking, one can distinguish two stages: (a) an almost linear increase of the length of the crack; and (b) an almost exponentially catastrophic increment of the length leading directly to destruction. It is a characteristic feature of the majority of wear processes, and even of the change of amplitude of machine vibrations in the wear process.

Each increment of the crack length (Equation (2.16) and Fig. 2.16) is the source of elastic wave (i.e. the acoustical emission) and the number of acoustic emission events per cycle:

$$\frac{d\Sigma}{dN} = C_2(\Delta K)^{s_2} \tag{2.11(a)}$$

Since K as well as ΔK is the function of the crack length (Equation (2.12)) leading to its almost exponential increment (Equation (2.16) and Fig. 2.16) a similar course in the function of the number cycles will have the summary number of sums Σ in the function of the number of fatigue load cycles (Fig. 2.17). It is easy to check qualitatively by substituting the estimated number of acoustical emission impulses in Equation (2.9) to the relation in Equation (2.11(a)). Therefore acoustical emission is a good indication and representation of the process of fatigue destruction.

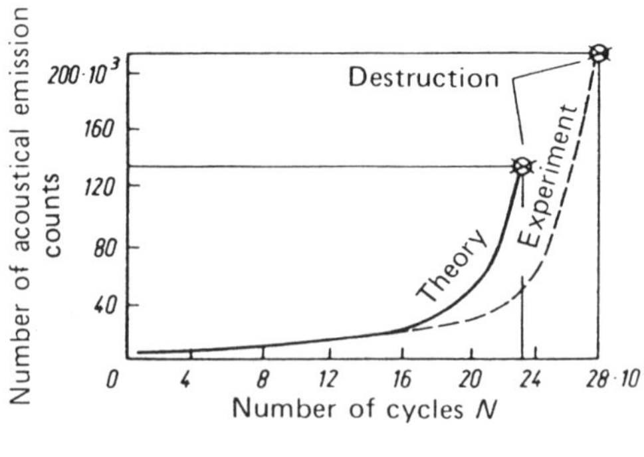

Fig. 2.17 — Comparison of the theory with the experiment of acoustical emission sums during fatigue test [36]

In conclusion, we can state that by watching the low-frequency vibrations we get the evaluation of the causes of fatigue; on the other hand, by watching the vibroacoustical process from the opposite end of the Figure 2.14 (i.e. acoustical emission) we have the evaluation of the results of fatigue destruction.

It is interesting to note the successful attempts of the evaluation of vibrational energy accumulated by a material sample in the occurrence of fatigue destruction. According to Tshatinian [42], the energy necessary for the destruction W_f at the given amplitude of stress σ_f and the number of cycles N_f can be estimated from Equation (2.17):

$$W_f = k \sigma_f^2 N_f \tag{2.17}$$

where k is the proportionality factor depending on the type of material. Thus, there is already a certain basis for the estimation of the *residual strength* of material which can be well investigated with the help of acoustical emission. Returning to the main topic of the phenomenon of fatigue, it must be emphasized that, so far, we have been talking only about the fatigue taking place in the volume of the material, while equally frequently the process of fatigue has primarily a surface character, e.g. in toothed gears and rolling element bearings where dynamic loads of the *surface layer* of the element are dominating. This type of *surface fatigue* is called *pitting* [29], and its nature is qualitatively similar to volumetric fatigue with modifications depending on the condition of the surface layer and the type of the lubricating agent. A general regularity is presented in the empirical Lundgren-Palagren relationship for rolling bearings:

$$p^3 N_f = \text{const} \tag{2.18}$$

where p is the unitary pressure and N_f is the number of loading cycles to destruction.

Summing up the discussion of fatigue wear, we can state that its intensity is directly proportional to the working load σ_m and the velocity of vibrations of element v. Fatigue increases microdamage and gaps which can be observed with the help of acoustical emission, since acoustical emission is a good indicator of fatigue.

2.3.3 Frictional wear

This form of wear takes place in kinematic pairs of machines which are characterized by mass decrements, and for their evaluation in general the thickness h of the worn layer is used, or the volume U of the worn material. In investigations and application considerations we use the theory of wear by microtruncation in the stationary version discussed above (see point 2.2.10), where the wear speed γ is constant. In such cases, the thickness of the worn layer in the function of the time of utilization is expressed by the relations [43]:

$$h = \gamma \cdot \theta = k\, p v \theta = k\, p s \tag{2.19}$$

where h is the thickness of abraded layer, γ is the speed of wear, θ is the time of utilization, k is a coefficient, p is the unitary pressure in the kinematic pair, v is the relative velocity of the pair of elements and s is distance.

For the volumetric wear, Archard's formula [44, 43] is appropriate:

$$U = \frac{k_0 p}{R_e} \cdot v \cdot \theta = \frac{k_0 \cdot p s}{R_e} \tag{2.20}$$

where the friction wear distinctly depends on the unitary pressure quotient p to the limit of plasticity of the material R_e, here k_0, similarly as k, in Equation (2.19), is a certain coefficient, depending on the conditions and the material of the kinematic pair, type of lubricant, etc.

In spite of much research effort in the field of friction wear there is no recognition concerning the influence of vibrations on the speed of wear. When we assume the simplest kinematic pair, like journal-bushing with relative vibrations in service conditions, we must take into consideration the dynamic component $p_d(t)$ of pressure in Equation (2.19):

$$h = k\,pv\,\theta = k\left[p_0 + p_d(t)\right]v \cdot \theta \tag{2.21}$$

where p_0 the stationary part of unitary pressure and $p_d(t)$ is the dynamic part of pressure with an amplitude proportional to the amplitude of relative vibrations of the journal bushing. We can see from Equation (2.21) a simple proportionality of mass (friction) wear to the amplitude of vibrations, which is often observed in the utilization practice for the so called *linear wear–normal wear*. However, if the vibration amplitudes exceed the limit value permissible for the given kinematic pair resulting from the clearance and the thickness of the lubricant layer, then the moments of a dry contact of the pair of elements and the speed of wear will grow considerably. This situation is presented graphically in Fig 2.18.

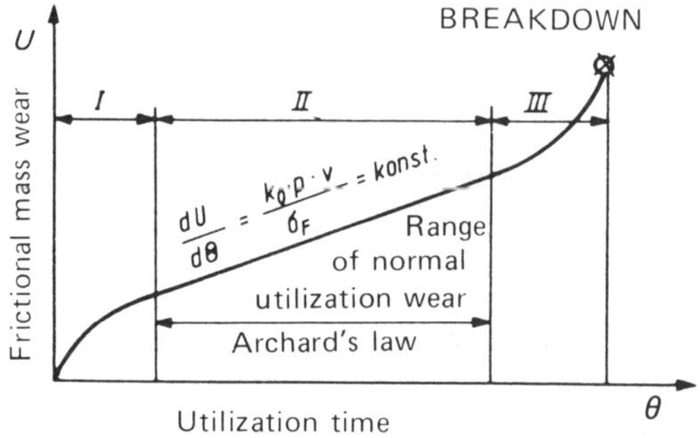

Fig. 2.18 — Stages of friction wear in kinematic pairs of machines

An important variation of friction wear is its combination with surface fatigue and corrosion, which is called *fretting*. Fretting takes place in the joints of machine elements which show relative microdisplacements (rivets, pins, keys and finger couplings) and even in kinematic pairs (bearings, gears) in the presence of high friction (generally in dry condition) and a corrosive atmosphere. The removal of one of these factors usually eliminates the process of wear by fretting. The fretting products consist of oxides on the one hand, and of surface wear in the form of microcracks and gaps on the other. Volumetric decrement of material U due to fretting can be expressed by an appropriate formula [29]:

$$U = \left[\frac{k_0(\sqrt{p}) - k_1 p}{f} + k_2 A\, p\right] N \tag{2.22}$$

where k_0, k_1, k_2 are constant factors, p is the unitary pressure, A is the amplitude of displacement of the relative vibrations in joints, f is the frequency of vibrations and N is the number of cycles of vibration. It can be seen that the vibrations (both amplitude and frequency) are factors determining the intensity of the wear process. And for constant frequency f the mass decrement depends linearly on the combination pressure p, the amplitude of the relative motion A, and the number of vibration cycles. This dependence is illustrated in Fig 2.19 [29].

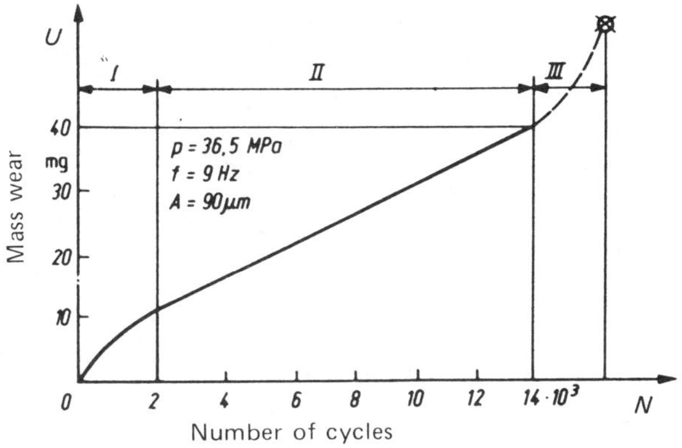

Fig. 2.19 — Exemplary mass wear in fretting of soft steel in a nitrogen atmosphere [29]

Also, two development stages of the fretting phenomenon are visible, and we can guess stage III which has been symbolically marked in the figure. In the first stage, the germination of fretting takes place; the second stage can be called the linear utilization wear since the number of cycles N is in a simple relation to the time of utilization θ. Finally, the third stage brings a sudden loss of the functional properties of the joint or the kinematic pair.

Summing up the above conclusions concerning the frictional wear and the influence of vibrations, we can say that for friction wear, vibration is a factor which accelerates the wear process, while for fretting, vibration is a factor conditioning this form of wear. In both cases, vibrations can play a role of the indicator of wear intensity of the frictional pair or the kinematic pair.

2.3.4 Cavitation wear

Among other forms of machine wear [29. 31, 43, 44] *cavitation erosion* can be essential for the problems of vibroacoustical diagnostics of machines and, particularly, of installations. Cavitation consists of the generation of vapour bubbles in the presence of decreased pressure of a fluid stream. Any sudden changes in the cross-section, direction of the stream and the motion of weak streamlines machine elements are the causes of cavitation phenomenon and cavitation wear. The wear develops as a result of bombardment of the surfaces of the element by shock waves created at the collapse of the cavitation bubble. Thus, there is a different wear in brittle materials giving craters and cavitational pinholes [29], from that of wear of plastic materials giving a characteristic corrugation connected with the flow of material. Generally, cavitation wear, particularly that of brittle materials, is inversely proportional to their hardness and directly proportional on the velocity of the fluid in the stream. The volume of the worn material can be expressed as [29]:

$$U = Bv^{n(\theta)} \cdot \theta \tag{2.23}$$

where $n(\theta)$ is the cavitation index, n depending on the time) is $n = 4 + 10$, B is the certain dimensional factor and v is the velocity of the fluid stream.

An illustration of Equation (2.23) is presented in Fig. 2.20 [29].

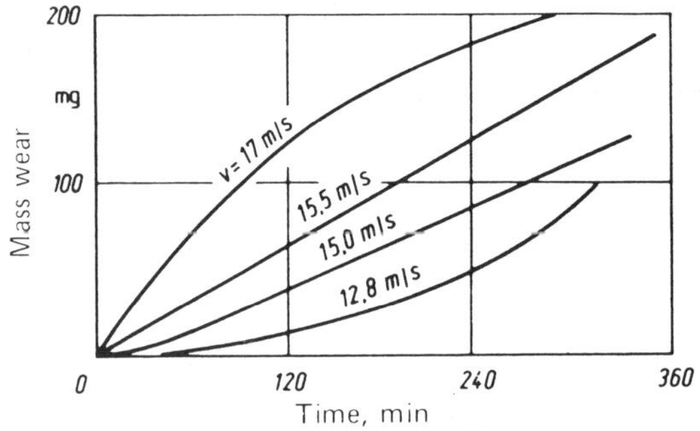

Fig. 2.20 — Cavitation wear as a function of time θ and velocity of fluid streamflow [29]

As mentioned earlier, cavitation causes strong high-frequency vibrations of the fluid and the construction elements. On the other hand, low-frequency vibrations of construction elements, particularly with high amplitudes, can themselves be the causes of cavitation. Thus, the observation of high-frequency vibrations makes it possible to estimate the intensity of cavitation in the installation, and the minimization of low-frequency vibrations of pipelines, etc. permits control to some degree of the creation of this unfavourable phenomenon.

2.3.5 Changes of form—creep

The creep of material and the changes in the form of machine elements connected with it are not counted as changes caused by wear. However, the effects which material creep evokes in the stressed machine elements working at high temperatures, and the influence of vibrations, fully justify regarding creep as a separate and significant form of wear. Creep can easily be defined by the example of simple stretching of a sample in a temperature of several hundred degrees centigrade. After the application of stress σ we obtain immediate elastic deformation $\varepsilon_0 = \sigma/E$. The constant force causes a slow increment of relative elongation, accelerated in the first period, stabilizes $\dot{\varepsilon} = (d\varepsilon/d\theta) =$ const in the second period, and accelerates and leading to failure in the third period, as shown in Fig. 2.21.

The simplest estimation of the deformation during creep can be presented in the following form [38, Chapter 19]:

$$\varepsilon = \varepsilon_0 + e \cdot \sigma^d \cdot \theta = \frac{\sigma}{E} + e \cdot \sigma^d \cdot \theta \tag{2.24}$$

where the constants e and d depend strongly on the material and temperature, while θ is the time of exposure.

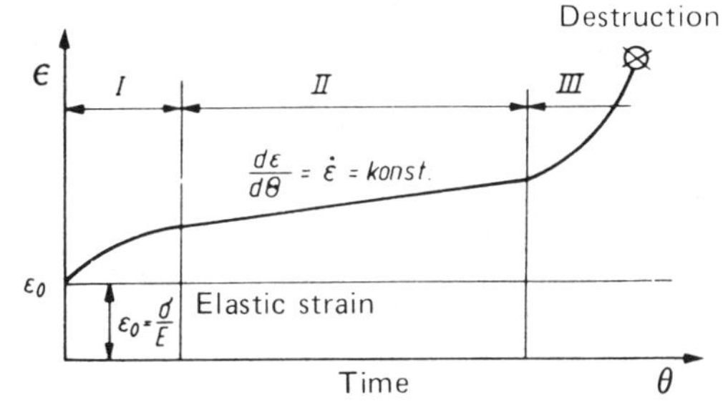

Fig. 2.21 — Three stages of the process of material creep in increased temperatures [38]

The present explanations of the phenomenon of creep are based on the dislocation movements which obtain additional activation energy from the applied stress and the increased temperature. The dislocation movement or the propagation of the crack can follow if the stress on the edge of dislocation exceeds the limits of plasticity which, for the given stress, decreases strongly with the increase of temperature.

However, as it is easy to observe, the temperature does not have to be the only activation source for dislocation and microcracks. It can be achieved by small vibrations superimposed on the working loads when $\sigma = \sigma_0 + \sigma_a \cos(2\pi ft)$). The fact that it is an important phenomenon is proved by Jakowluk and other work on creep and vibrocreep [45].

For high frequencies of vibrations of the order of several tens of kilohertz superimposed on the original load σ_0, the decrease of plasticity limit and the possibility of creep of material is significant, even in room temperature. It can lead even to an 80% decrease of the value of plasticity limit [46]. These results are illustrated by the diagrams of tension [46] and compression of samples [47] with different amplitudes of ultrasound vibrations (Fig. 2.22).

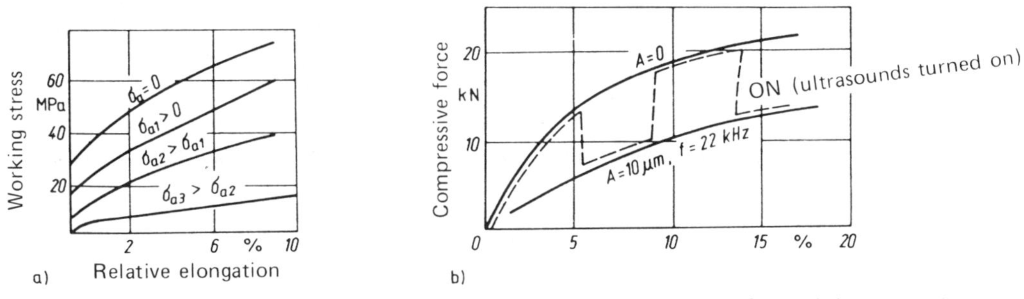

Fig. 2.22 — The influence of ultrasound on the process of tension of aluminium samples [46] (a) and the compression of the same material [47] (b) with a visible effect of ultrasound switched on and off (broken line)

When we look at the relation of creep in Equation (2.24) in the light of these results, we must state that the second member of the relation must depend strongly on temperature and the frequency of vibrations (or frequency of the dynamic component of stress). Having this in mind, for qualitative purposes, we can write this relation as:

$$\varepsilon = \frac{\sigma_0}{E} + \mathit{c}(T)\,\sigma_0^d \cdot \theta + g(f)\,\sigma_a \cdot \theta \qquad (2.25)$$

where T is the sample temperature, f is the frequency of the dynamic component of stress, g is a constant depending on the type of material and the conditions of testing and θ is duration of test. The latter presents vibrocreep and it becomes unusually significant for the frequency of the order of several kilohertz.

Summing up this section, we must state that vibrations, particularly those of high-frequency, can considerably accelerate this process at high temperatures, and at low temperatures they can promote creep and change of form.

2.3.6 Tribovibroacoustical model of a machine

There is an interrelation between vibration and wear detected experimentally [7]. However, there is no mathematical model of a machine combining together the vibroacoustical and wear phenomena, or tribological phenomena, since all the wear phenomena of machines are dealt with by tribology [30, 31, 43]. The first, partially successful, attempts to outline a tribovibroacoustical model were undertaken by the Author in 1980 [48]. A better version of this model is presented below. It is based on the stated assumption of the proportionality of vibrations D to the advancement of the wear processes. This advancement is best measured by energy lost for the irreversible wear processes—tribological processes E_z. Remembered that a new machine (without wear) is characterized by a certain level of rated vibrations D_n we can write our relation of diagnostics in the following form:

$$D = D_n + \gamma E_z = D_n + \gamma N_z \cdot \theta \tag{2.26}$$

where D is the amplitude measure of vibrations (e.g. acceleration, velocity, etc.), $D_n = D\ (\theta = 0)$ is the level of rated vibrations of a new machine, E_z is the energy lost for irreversible wear processes, γ is the coefficient of tribovibroacoustical loss presenting a linkage between the tribological processes and vibrations, and N_z is the mean power lost for tribovibroacoustical processes. The latest much improved version of the tribovibro-acoustical machine model was published by the Author as "Generalized tribovibro-acoustical (TVA) model of the machine in plant diagnostic environment", *Bulletin of the Polish Academy of Sciences*, Technical Sc., 1990, Vol. 38, No. 1–12.

The lost energy E_z can be estimated very easily knowing the mechanical efficiency of the machine η since:

$$\eta = \frac{N_{u\dot{z}}}{N_{za\acute{s}}} = \frac{N_{za\acute{s}} - N_z}{N_{za\acute{s}}} = \frac{E_{za\acute{s}} - E_z}{E_{za\acute{s}}} \tag{2.27}$$

where $N_{u\dot{z}}$ is useful power, $N_{za\acute{s}}$ is supply power and E is adequate energy.

The tribological processes which most frequently accompany the motion of machines are fatigue and friction, and the energy released from them will be proportional to the length of the fatigue gap l and to the volume of the worn off material U. We will limit our considerations to two phenomena. Let us assume:

$$E_z = E_{\text{friction}} + E_{\text{fatigue}} + \ldots \cong al + bU \tag{2.28}$$

where a and b are certain coefficients of proportionality, l is to be calculated from the Paris formula (Equation (2.16)) and U from Archard's formula (Equation (2.20)). Calculating the length of the fatigue gap by the integration of Equation (2.16), in Kocańda's version [40] we find:

$$l = c_3\left(\sigma_0+\sigma_a\right)^n N = c_3\left(\sigma_0+\sigma_a\right)^n \cdot f \cdot \theta$$
$$= c_3\left(\sigma_0 + eD\right)^n \cdot f \cdot \theta \tag{2.29}$$

where e is the coefficient of proportionality of vibrations D to variable stresses σ_a, f is the frequency of vibrations and θ is the time of their action.

By inserting the dynamic component of pressure $p = p_0 + gD$ into Archard's formula (Equation (2.20)) and including the coefficient of proportionality to the amplitude of vibrations g, we have:

$$U = \frac{k_0 c \theta}{R_e}\left(p_0 + g D\right) \tag{2.30}$$

Substituting Equations (2.30) and (2.29) into the expression for energy of utilizational processes E_z given in Equation (2.26) we will have:

$$D = D_n + \gamma\left[\alpha c_3\left(\sigma_0 + eD\right)\right]^n \cdot f + \frac{bk_0 v}{R_e}\left(p_0 + g D\right)\cdot \theta = D_n + \gamma N_z \theta;$$
$$N_z = \alpha c_3\left(\sigma_0 + eD\right)^n \cdot f + \frac{bk_0 v}{R_e}\left(p_0 + g D\right) \tag{2.31}$$

where we have introduced in an obvious way the power of wear processes $N_z = E_z / \theta$ which, as we can see, also depends on the amplitude of vibrations D. Utilizing this dependence for the final determination of the amplitude of machine vibrations in the function of its utilization time θ, from Equation (2.31) we can find:

$$D = \frac{\gamma N_z^0 \cdot \theta}{1 - \gamma\theta \left.\dfrac{dN_z}{dD}\right|_{D = D_n}} + D_n \tag{2.32}$$

where $N_z^0 = N_z\,(D = D_n)$ according to Equation (2.31), and dN_z/dD is a gradient of the power of the wear processes due to the vibrations for a hypothetical new machine (i.e. $D = D_n$); mathematically it is the coefficient of N_z development into Taylor's series in relation to vibrations D presenting the sensibility of the power of wear processes to vibrations.

Analysing the final formula of our tribovibroacoustical (TVA) model Equation (2.32), we must distinguish in it two characteristic points. For time $\theta = 0$ we obtain $D = D_n$ i.e. the amplitude of rated vibrations of the new object.

On the other hand, for $\theta = \theta_{aw} = \left(\gamma \left.\dfrac{dN_z}{dD}\right|_{D = D_n}\right)^{-1}$ we find $D = \infty$,

which can be interpreted according to experience as a breakdown. The value of the time to breakdown:

$$\theta_{aw} = \left(\left.\frac{dN_z}{dD}\right|_{D = D_n} \cdot \gamma\right)^{-1} \tag{2.33}$$

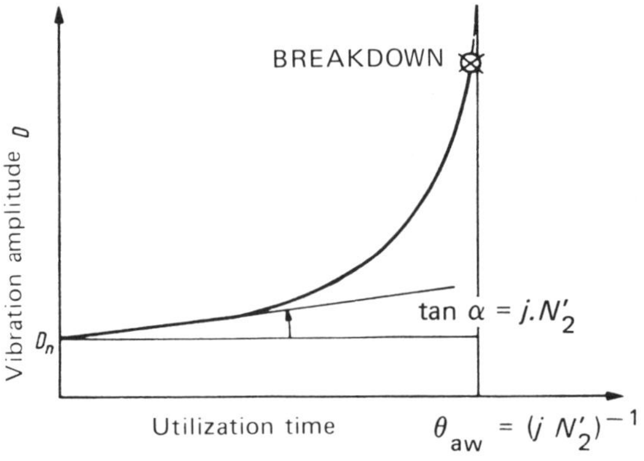

Fig. 2.23 — Graphic presentation of vibration amplitude in the model of a tribovibroacoustical machine

is inversely proportional to the tribovibroacoustical loss factor γ and the sensitivity of wear to vibrations dN_z/dD. On these characteristics and on the combined power of tribological processes N_z^0 of a new machine, depend both on the value of the amplitude of vibrations D and the velocity of its changes in time of utilization θ. The qualitative picture of these changes is presented in Fig. 2.23, where we see that both the time of breakdown θ_{aw} and the speed of the increase of the vibrational symptom ($tg\alpha$) are functions of tribovibroacoustical loss factor γ, tribovibroacoustical sensitivity dN_z / dD and the combined power of the wear process N_z. These are not common conclusions; hence, it seems that the model itself is very valuable. This model indicates that the time of breakdown θ_{aw} can be predicted for a new machine when we know only the vibrational gradient of the lost power dN_z / dD and its tribovibroacoustical loss factor γ. Also it is possible to utilize the model and Equation (2.32) in sequence for the prediction of vibrations and the condition of machines; this is the major topic of the Author's investigations (see Section 4.5.2).

To conclude our considerations on the tribovibroacoustical linkage, let us look once again at Fig. 2.14 from the point of view of causes and symptoms of wear in mutual feedback, as in Fig. 2.24. We can see from the figure that the primary motion actions are the reasons of low-frequency vibrations which cause fatigue microdamage, etc. The damage accumulates during the utilization process giving indirect damage and microfailures it increase the level of the original dynamic actions. This, in turn, increases the level of microdamage in the mutual feedback leading, in effect, to the acceleration of wear and breakdown.

2.4 SUMMARY

We can state that, by the analysis of the dynamic and wear phenomena, we have succeeded in creating a tribovibroacoustical model of a machine which gives the basis for diagnostic observation and prediction of vibrations, and have also explained the

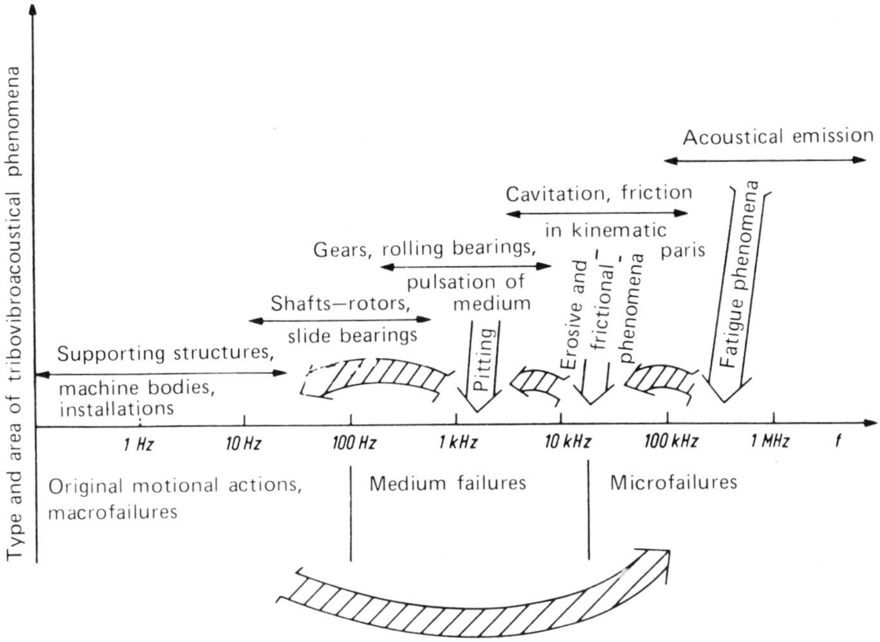

Fig. 2.24 — Vibroacoustical and tribological phenomena in machine in mutual feedback

frequency division of the particular wear phenomena. In the field of low frequencies, as shown in Fig. 2.24, all wear effects accumulate, while for high frequencies there is only acoustical emission as the symptom of fatigue wear. This spectral decomposition of the wear effects will prove useful in diagnostics.

3

Failure-oriented processes, signals and measures of vibroacoustical phenomena

3.1 INTRODUCTION

Bearing in mind the discussion in the previous chapters, we can identify the most important vibroacoustical processes generated by operating machines and what kind of important wear processes occur in them. We have also clarified the general relations between the vibroacoustical and the tribological processes and we have arrived at the general tribovibroacoustical model. However, we are aware that the majority of these tribovibroacoustical processes in machines take place at the same time in different subassemblies and kinematic pairs. Thus, the total vibroacoustical process obtained in any part of a machine's construction is at least the sum of vibroacoustical processes taking place at the same time in the machine system. Therefore, it is necessary to study the possible observation methods of vibroacoustical processes in machines, in order to be able to extract from that great amount of information the vibroacoustical process we are interested in and what is the measure of the occurring wear process. Furthermore, the proper *measures of signals* must be applied to this vibroacoustical process transformed by a transducer into a vibroacoustical signal, so that the *symptom* obtained in this way is a vibroacoustical symptom of wear taking place in the machine, a symptom giving evidence of the intensity of wear and its advancement. Then a complete diagnosis of the machine is possible, i.e. a diagnosis of the condition, its origin, and even prognosis or the estimation of the time to breakdown.

3.2 OBSERVATION MODELS OF TRIBOVIBROACOUSTICAL PHENOMENA IN MACHINES

So far, we have discussed wear processes in the model aspect in the investigations of elementary single phenomena. In real objects the wear processes lead to definite failures with different degrees of advancement and if it is not stopped in time complete breakdown occurs. Furthermore, the same type of wear occurs in many machine elements with a different intensity (e.g. slow-speed bearings get worn out in a different way than high-speed ones) and the fatigue process takes place in the blade of a turbine and in its shaft, but the latter happens in a different way and is rarely observable. From the above we can conclude that it is better and even necessary to speak about the wearing out of a definite element or subassembly, which we call *failure*. Thus, the wear of an element equals a gradual failure taking place during the life of an object θ.

3.2.1 Failure life

Now let us introduce a summary measure of the advancement of a given failure $F(\theta)$ that is called the *curve of failure life* and presents the development of the failure from the moment of its activation $\theta = 0$ until the moment of its regeneration θ_r. For example, $F(\theta)$ can be dimensional or nondimensional measure of utilization unbalance of an exhaust fan in a power unit between two successive stoppages for the balancing of the rotor. (Fig. 3.1).

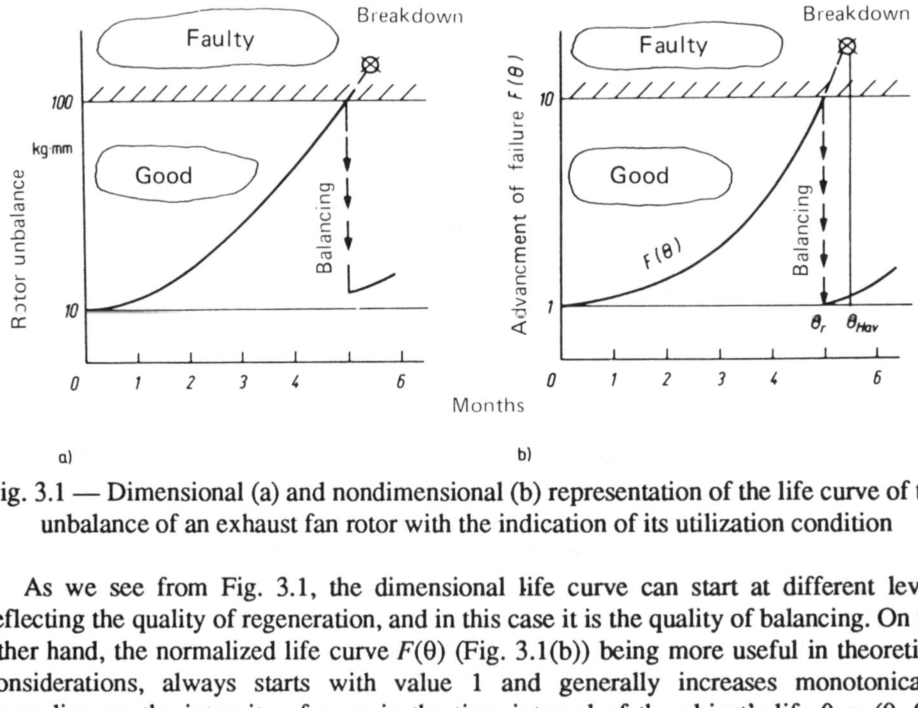

Fig. 3.1 — Dimensional (a) and nondimensional (b) representation of the life curve of the unbalance of an exhaust fan rotor with the indication of its utilization condition

As we see from Fig. 3.1, the dimensional life curve can start at different levels reflecting the quality of regeneration, and in this case it is the quality of balancing. On the other hand, the normalized life curve $F(\theta)$ (Fig. 3.1(b)) being more useful in theoretical considerations, always starts with value 1 and generally increases monotonically depending on the intensity of wear in the time interval of the object's life $\theta \in (0, \theta_r)$. Both curves end at the moment the limit of utilization efficiency is transgressed, while in exceptional cases they end at the moment of a breakdown θ_{aw}. So we can accept the rule that the value of the dimensional life curve is the direct measure of the condition of the object, and it can be compared with the limit value F_{gr}. So the assessment of the condition will consist of the comparison of $F(\theta) \lesssim F_{gr}$, and in a two-class evaluation of the technical condition we will have:

$$F(\theta) < F_{gr}, \qquad X = z = \text{Good Condition}$$
$$F(\theta) > F_{gr}, \qquad X = n = \text{Faulty Condition} \tag{3.1}$$

In a real mechanical object we have to deal with more than one potential failure. The advancement of these failures will be represented by a family of life curves $F_i(\theta)$, $i = 1, \ldots, n$, and each of them should have its vibroacoustical equivalent discriminating one curve among many curves, or one failure among many failures with different character and intensity. It is easy to guess that it is not a simple task; therefore, we are going to develop a special model of diagnostic observation of an object with *n* life curves, or an object with *n* dimensional *space of failures*. Then we can say that the *condition of machine* X is determined not by one number ($F(\theta)$ value) but by a set of numbers, or that it is a vector $\{\mathbf{X}\} = \{\mathbf{x}_1, \mathbf{x}_2, \mathbf{x}_3, \ldots, \mathbf{x}_n\}$, and in a particular case it is the *vector of the condition* of the form: $\{\mathbf{X}\} = \{\mathbf{F}_1(\theta), \ldots, \mathbf{F}_n(\theta)\}$.

Concluding the introduction of the notion of life curve presenting the change of the condition of the object, it is worth clarifying the problem of the independence of life curves: $F_1(\theta), \ldots, F_n(\theta)$. Namely, the observation of systems subject to gradual degradation including mechanical systems and biological systems leads to the conclusion that we can speak about the independence of particular failures $F_i(\theta)$ only in a new system. With the wear of the object, the particular life curves become increasingly less independent and deterioration of the condition of one element leads to the deterioration of the condition of other elements. By analogy to man we can speak about a *syndrome of an old man* [7, p. 306]; here, the internal organs and processes taking place in them, originally independent, start deviating increasingly and simultaneously from the standard, leading in effect to death. In the mathematical way, one can express this in the form of a small correction to each life curve making it dependent on other ways of damage by a small parameter $\varepsilon(\theta)$. The new life curve with the consideration of the syndrome of an old man, $\tilde{F}_i(\theta)$ can have the form:

$$\tilde{F}_i(\theta) = F_i(\theta) + \sum_{i=1}^{n} \varepsilon_j(\theta)\, F_j(\theta)$$

$$\varepsilon_j(0) = 0 \qquad\qquad \theta \leq \varepsilon_j(\theta) \ll 1$$

where, as we can see, a great number of small influences expressed by $\varepsilon_j(\theta) \ll 1$ can have a significant meaning. We are not going to use or develop this conception further, but it is worth keeping it in mind when we frequently encounter a disagreement between the diagnostic facts and the applied simplifications of phenomena models.

3.2.2 Model of diagnostic observation of an object—vibroacoustical signal

Let us consider the construction of a machine which can be modelled by a mechanical system with a continuous distribution of mass, stiffness and damping (the coordinate *r*) in which such tribovibroacoustical phenomena occur like volumetric fatigue, surface fatigue, fretting, frictional fatigue, etc. These phenomena are connected by the movable and immovable elements and subassemblies of the machine. Let us mark the life curves of these elements by $F_i(\theta)$, $i = 1, \ldots, n$. These failures, each one separately, are the sources of the original vibroacoustical process, i.e. $V_i(t, \theta)$, which spreads in the form of vibrations throughout the machine structure being subject to time transformations '*t*' and spectral transformations '*f*' by its multiresonance characteristics. The original cause of the motion, vibrations and other tribovibroacoustical phenomena in the machine system are the forced motions such as rotation connected with the machine operation. These forced motions give the so-called rated vibrations $c(r, t, \theta)$ whose level and spectral composition depend on the machine type. For example, the rated vibrations of a piston compressor are diametrically different from those of a rotor compressor. This sum of forced motions and vibroacoustical reactions of the construction give, in a certain position r, the vibroacoustical process we are going to denote by $A\ (r,\ t,\ \theta)$. In this position a process-signal transducer (e.g. a transducer of vibration acceleration) is located, and on its output is recorded the electric signal $s\ (r,\ t,\ \theta)$. The model of diagnostic observation of a machine outlined above is presented in Fig. 3.2.

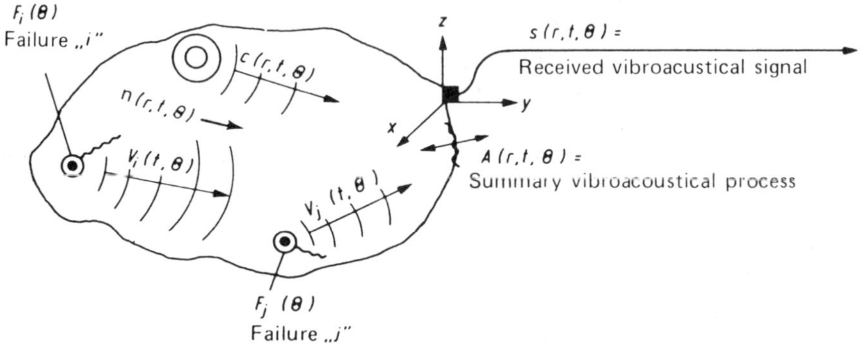

Fig. 3.2 — Model of diagnostic observation of a machine body

In order to obtain an analytical model of the received signal $s\ (r,\ t,\ \theta)$ we are going to define the impulse functions of the transition from the i-th place of failure to the point of reception, i.e. $h\ (r_i,\ t,\ \theta)$, where $r_i = r\ (x_i,\ y_i,\ z_i)$ is the vector leading from the place of reception to the sphere of vibroacoustical process generation. As we know, the construction response (generally the response of the mechanical system) to an optional forced motion is a convolution of stimulation $V_i\ (t,\ \theta)$ and impulse function of the system $h\ (r_i,\ t,\ \theta)$. Thus, the signal reproduced by the transducer in the beginning of the system of coordinates according to the above and Fig. 3.2 will be:

$$
\begin{aligned}
s(r,t,\theta) &= \sum_{i=1}^{n} \int_{0} h(r_i, t-\tau, \theta) V_i(\tau,\theta)\, d\tau + c(r,t,\theta) + n(r,t,\theta) \\
&= \sum_{i=1}^{n} h(r,t,\theta) * V_i(t,\theta) + c(r,t,\theta) + n(r,t,\theta)
\end{aligned}
\tag{3.2}
$$

where the asterisk indicates the operator of the convolution in reference to dynamic time t, $c\ (r,\ t,\ \theta)$ is the rated signal received by the transducer, and $n\ (r,\ t,\ \theta)$ is the noise inherently connected with each observation and cognitive activity of man. The interpretation of this noise component can be very flexible here, ranging between noise incoherent with any original failure signal $V_i\ (t,\ \theta)$ and a case where all failure signals except the investigated one are interfering noises. This example shows the subjectivity of the notion of useful information and noise, a variable depending on the reason for diagnosis.

3.2.3 Measures of vibroacoustical signals—symptoms and discriminants

As has been emphasized, diagnostics means in the first place the determination of the technical condition (for the time being this condition is coded in the quickly changing vibroacoustical signal = $s(r, t, \theta)$). As we can see, it has three fields of definiteness, r, t, θ, and due to the random noise component $n(r, t, \theta)$, the process (signal) should be regarded in each field as a random one. In the field of space r it is a process with a finite mean value and finite variance, and in the field of dynamic time t it is a stationary random process with zero mean value and finite variance. On the other hand, in the field of utilization time θ it is a random growth process (linear or nonlinear regression), reflecting the summary character of wear (degradation) of the object. The above indicates that in order to represent the wear and its evaluation (i.e. in order to define the condition) we must apply to the signal in Equation (3.2) an adequate measure or estimate. Generally [49], we can distinguish two types of signal measures—*symptom* and *discriminants*—although it is better to assume that the discriminant is a particular type of a symptom. Symptom is a measure of a general type defined as in Equation (3.3):

$$\begin{aligned} S(r,\theta) &= E_t\{\Phi(s)\} = \frac{1}{T}\int_0^T [\Phi s(r,t,\theta)]\, dt \\ &= \Phi_0\left[\Sigma a_i(r)V_i(\theta) + c(r,\theta)\right] + N(r,\theta) \\ v_i(\theta) &= E_t\left\{V_i(t,\theta)\right\}, \qquad i = 1,\dots,n \end{aligned} \tag{3.3}$$

where $E_t(\cdot) = \frac{1}{T}\int_0^T (\cdot)\, dt$ is the operator of averaging in respect to dynamic time t with the time of observation (averaging) T, while $\Phi(\cdot)$ is the operation carried out on the signal to extract information, and operator $\Phi_0(\cdot)$ indicates a functional dependence of the symptom on the indicated arguments with the function of mass $a(r)$ instead of the operation of convolution, and where $N(r, \theta)$ is a random disturbance of the symptom depending on r and θ.

As we can see from the above, the symptom of a vibroacoustical diagnostic signal according to Equation (3.3) is the representation of all wear processes in the object, i.e. it speaks about the general condition of the object. In practical investigations of vibroacoustical diagnostics, a similar role for all machines is played by the measurement of the peak value of vibration velocity while the measurement of the peak acceleration of the machine body plays the same role for industrial toothed gears, etc. for details, see Section 3.5 and the Author's monograph [7].

On the other hand, the discriminant has properties selective of the total number of original vibroacoustical signals of failure (e.g. the k-th signal according to the general relation shown in Equation (3.4)):

$$\begin{aligned} D_k(r,\theta) &= E_t\{\Phi(s)\} = \frac{1}{T}\int_0^T \psi[s((r,t,\theta))]\, dt \\ &= \psi_0\left[a_k(r)V_k(\theta)\right] + N(r,\theta) \end{aligned} \tag{3.4}$$

where $E_t(\cdot)$ is the operator of averaging with respect to time, $\psi(\cdot)$ is the operator of signal transformation, $\psi(\cdot)$ shows only the functional dependence, while $N(r, \theta)$ is a random disturbance.

The simplest example of such a discriminant is the averaged amplitude of vibrations in the rotation band as the indicator of unbalance. More complicated is the kurtosis β as a measure of the flattening of distribution density of vibration amplitude in the application to the evaluation of the wear of rolling element bearings. This is discussed in detail in Section 3.5.3.

Having defined the measures of vibroacoustical signals we can come back to the main question of vibroacoustical diagnostics, i.e. to the vibroacoustical representation of wear advancement. In creating our model of diagnostic observation we have assumed that the life curve $F_i(\theta)$, being an illustration of the advancement of the i-th failure, model the original signal of failure $V_i(t, \theta)$, i.e. after averaging there exists an assumed complete proportionality:

$$F_i(\theta) \sim E_t\{V_i(t, \theta)\}, \qquad i = 1, \ldots, n \tag{3.5}$$

In the light of this assumption we can ascribe to Equation (3.3) the proportionality to the total advancement of the wear process:

$$S(r,\theta) = F_0\left[\sum_{i=1}^{n} a_i(r) V_i(\theta) + c(r,\theta)\right] + N(r,\theta) \simeq$$
$$= \sum_{i=1}^{n} a_i F_i(\theta) + N(r,0) \tag{3.6}$$

On the contrary according to Equations (3.4) and (3.5), this discriminant represents one k-th failure, since:

$$D_k(r,\theta) = \psi_0\left[a_k(r) V_k(\theta)\right] + N(r,\theta) \simeq a_k F_k(\theta) + N(r,\theta) \tag{3.7}$$

Having in mind the above, we can finally outline the sphere of action and the way of reasoning used in vibroacoustical diagnostics of machines. In other words, we could say that we can outline the model for vibroacoustical diagnostics of machines.

> During machine operation, the i-th failure whose advancement is reflected by the life curve $F_i(\theta)$ models its original vibroacoustical process $V_i(t, \theta)$. This process, spreading in a wave motion in the machine construction, and being mixed with other processes and disturbances, creates a resultant vibroacoustical process in the place of reception $A(r, t, \theta)$. This summary process transformed into a signal $s(r, t, \theta)$ is subsequently subject to a treatment leading to the creation of a measure and compression of disturbances giving, in effect, a vibroacoustical symptom of wear $S(r, \theta)$ from which, assuming a proportionality to wear of the type in Equations (3.6) and (3.7), we draw the final conclusions about the condition of the machine $X \, \varepsilon \, (z, n)$.

This model can be represented as a sequence of implication relations as in Fig. 3.3.

Taking into consideration the disturbance components of the symptoms in Equation (3.3) and discriminants in Equation (3.4), our choice of the diagnostic signal, the place of its reception 'r' (Fig. 3.3) and type of the applied symptom, should be such as to maximize the correctness of diagnostic inference. It is the main problem of machine vibroacoustical diagnostics not yet solved explicitly, since the final information is always an *estimation of the condition* and not information about the real condition of the object.

3.2.4 Observation area of vibroacoustical processes in machine diagnostics

Having in mind the above outlined model of vibroacoustical diagnostics, let us look at the application of vibroacoustical diagnostics in the life of a group of N machines of the same

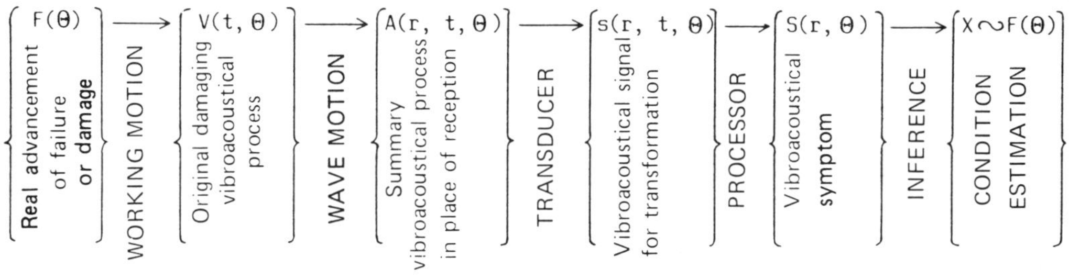

Fig. 3.3 — Basic cause-and-effect sequence used in vibroacoustical diagnostics of machines

type. For each individual machine we can measure signal s_j (r, t, θ), $j = 1,\ldots, M$. Let us note, however, that even with the same received vibroacoustical process (e.g. vibration acceleration) by changing the place and direction of measurement r in machine construction we have at our disposal a whole family of vibroacoustical processes. Investigating the similarity of processes within this family for the same time of life θ = const (particularly $\theta = 0$) we have a case of construction diagnostics, where we estimate the dynamic condition of the construction prototype. Here, the life curves for time $\theta = 0$, i.e. F_i (0) do not symbolize failure but an independent source of vibrations in machine construction whose identification and minimization is the subject of investigations in construction diagnostics.

Remaining still at the zero life time of the object $\theta = 0$, and investigating the same vibroacoustical symptom for different individual machines, we have the possibility of comparative examinations with the *quality standard* of a new machine. This type of investigation of the vibroacoustical processes for r = const $\theta = 0$, j = var is used in the investigation of quality in *checkout* diagnostics. Here, in turn, we identify the initial values of life curves F_{ij} (0) as the measure of quality which, within the investigated machine group j = var, are subject to a definite distribution.

If we now determine the observation place r = const for the same machine j = const, and we investigate the evolution of a definite measure (symptom) with a variable utilization time θ = var, then we will obtain information about the change in the utilization availability of the machine in relation to the *good condition standard* of a new machine. This is the domain of the utilization diagnostics (condition monitoring) where we identify the type of life curve (type of failure) F_i (θ) and its advancement. Instead of interpreting F_i (θ) as the life cure, we can also assume that it is a quantitative picture of the parameter determining the quality of the technological process. Then, as we see, for the same r = const, j = const, θ = var, we have to do with the diagnostics of manufacturing processes.

The fields of the application of vibroacoustical processes outlined above and our model of their observation can be expressed synthetically by defining a three-dimensional orthogonal area of observation and diagnostic application with axes of dynamic time t, utilization time θ, and the ordinary number of the object in the group j (Fig. 3.4).

As we see in Fig. 3.4, we can investigate the similarity and the *dynamic condition* of the given individual machine along axis t; along axis θ, we can investigate the similarity and its *utilization condition*; while along axis j, we can compare it with other objects and with the standard defining its *quality condition*.

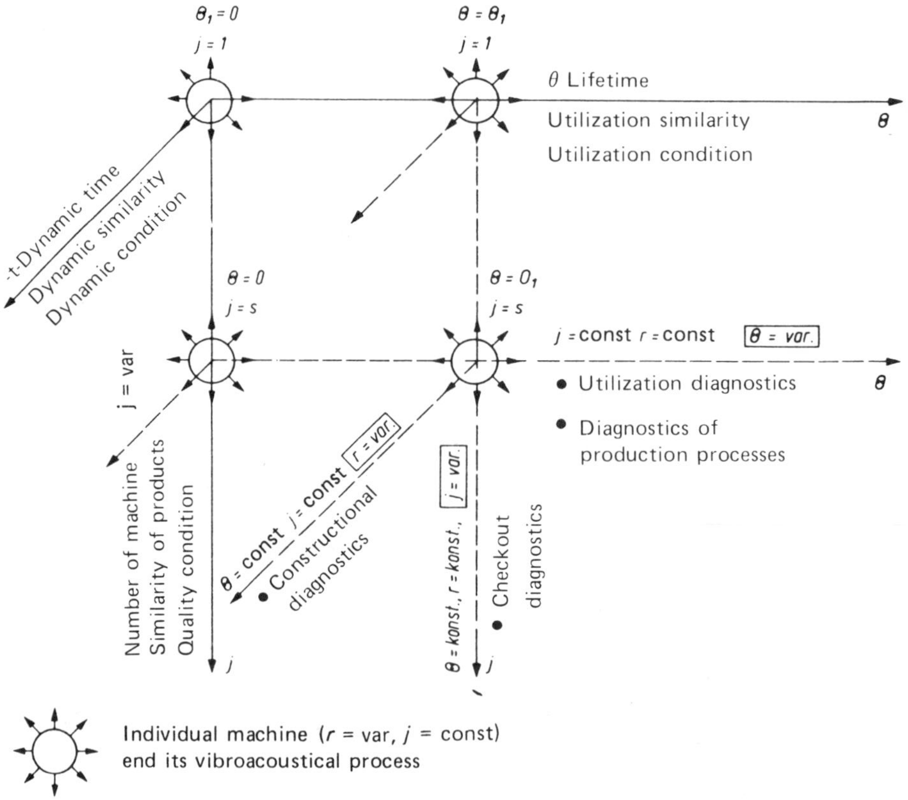

Fig. 3.4 — Application space (field)of the vibroacoustical processes in machine diagnostics [7]

3.3 SELECTION OF DIAGNOSTICALLY USEFUL VIBROACOUSTICAL PROCESSES

The first step in diagnostics is the determination of its detailed objectives following on from a clearly defined *diagnostic task*. It means that when we have to determine the condition of a machine, we have to define first the components of the vector of its condition $\{\mathbf{X}\} = \{\mathbf{x}_1,\ldots, \mathbf{x}_n\}$ (Section 3.2.1) or.in other words, we have to define the area of failures. Speaking in terms of diagnostic application, we have to determine the potentially possible machine failures, their probability of occurrence and the mean failure-free working time for each type of failure, etc. It is desirable to support these data by reliability and economic analysis as recommended in Chapter 8 of the Authors monograph [7]. Such formulation of the problem by the selection of the dimensions of the vector of condition $\{\mathbf{X}\}$ is the result of the valuation of needs, and is an attempt to answer the question: *What should be diagnosed in the given machine?* We will come back to this question in the examples of Chapter 5, for the time being we shall assume that the failure area of the machine if determined in such a way that the dimensions and character of the vector of the condition are known. It means we assume that we know what we need to know', which is a great achievement by itself. A further step in the elaboration of the problem the diagnostics of a mechanical object attempt to define the vibroacoustical symptoms of the condition being an analogue of the vector of condition $\{\mathbf{X}\}$. This problem will be discussed in Section 3.3.1

3.3.1 A tribological model of the generation and propagation of vibroacoustical phenomena

When we want to use the information about the technical condition contained in the vibroacoustical processes, we must first become acquainted with their mechanisms of generation and propagation. So let us take into consideration the experimental facts possible for identification in each machine. The analysis of dynamic processes such as vibrations, noise, pulsations, stresses, instantaneous driving moment, etc. in machines with a stationary periodic motion show that these processes can be regarded as weakly periodic. They are composed of a determined polyharmonic process and of disturbances with a wide spectrum, the so-called 'noise'. A detailed spectral analysis of these processes confirms this calculation, utilizing the presence in the spectrum of the basic frequency connected with the number of revolutions and a series of linearly dependent higher spectral components, e.g. harmonics. Furthermore, one can indicate types of machines and installations where we can determine in advance what component in the spectrum of the process is connected with the work of the given element of the machine. However, the amplitudes of these components are subject to strong fluctuations so that the prediction of the condition of the machine on the basis of the amplitude of the spectral component can be carried out independently, but only in simple cases. There are many causes of these fluctuations and the most important of them can be characterized in the following ways.

Firstly, the instantaneous frequency of the basic working motion of a machine is subject to small changes within one revolution and between the particular revolutions due to heterogeneity (e.g. impacts) and randomness of working loads. Therefore, the whole series of harmonics connected linearly with the revolutions are subject to the same fluctuations.

Secondly, the measurement of a diagnostic signal (usually vibrational signal) usually takes place at one of the points of the dynamic system of the machine. These systems have a mechano-acoustical character and are characterized by a multiresonance transmittance and small dynamic losses. Therefore, each significant fluctuation of the series of harmonics connected with the basic motion of the machine gives a strengthened fluctuation to the reaction of the machine structure considered as a dynamic system. In consequence, this fluctuation is an additional disturbance of the diagnostic signal received from a definite point in the machine structure. The above-mentioned small damping of the mechanical structure for the frequencies of the order of hertz and kilohertz is, furthermore, the reason for a great number of components of the vibroacoustical process originating from different remote places of the machine structure and having the character of disturbances from the point of view of the objective of diagnosis. From the point of view of the propagation of vibrations, the damping starts playing an essential role first for the frequencies of the order of 10–20 kHz or more, where no further resonance phenomena are observed and the amplitude of the elastic wave decreases with the distance r at least like $1 / r$.

Another cause of the fluctuation of vibroacoustical processes is the presence of wide-band noise with an additive or modulation character. This can be caused by frictional phenomena in kinematic pairs, acoustical emission, external causes and, generally, by the tribovibroacoustical phenomena which are not actually the objective of diagnosis.

The dynamic components of the working process and tribovibroacoustical processes are subject to the evolution of amplitude and spectrum with the progressing wear which among others reflects the influence of the utilization time θ. Assuming the period of the basic motion of the machine is T, following the Author's example [50], we can proceed to the construction of a tribovibroacoustical model of the generation of vibroacoustical phenomena in machines with a periodical motion. The most important characteristics of this model are specified as follows.

(1) The condition of the machine (mechanism, subassembly) is uniquely defined by the characteristic signal φ_i (t, θ) $0 \le t \le T$; $0 \le \theta \le \theta_{aw}$, generated separately with each revolution, $i = 1, \ldots,$. This signal, also called an elementary event, is subject to changes in the quick dynamic time t and to evolution in the slow time of the object's life θ.

(2) The characteristic signal consists of the determined process φ_0 $(\cdot)$ and the random process n_i $(\cdot)$, while their intensity and spectral composition characterizes the condition of the object. Thus, during the i-th revolution the elementary event is generated:

$$\varphi_i(t,\theta) = \varphi_0(t,\theta) + n_i(t,\theta), \quad t\,\varepsilon(iT,(i+1)T) \;\; i=1,\ldots \tag{3.8}$$

(3) The rotational motion of the machine (generally periodic) transforms the characteristic signal by the so-called kinematic transformation into a continuous signal x (t, θ). In the first approximation, this transformation can be understood as T-periodic extension (repetition), i.e.:

$$x(t,\theta) = \sum_{i=1} \varphi_i(t - iT)\, U(t - iT) - \sum_{i=1} \varphi_i(t,0) * \delta(t \quad iT) \tag{3.9}$$

where $U(q) = \begin{cases} 0, & q \le 0 \\ 1, & q > 0 \end{cases}$ while the asterisk denotes the already known operation of convolution and $\delta(t - iT)$ is the period Dirac's distribution.

(4) The characteristic signal transformed in this way and being a dynamic representation of the condition of the object, forces the vibrations of the surrounding mechano-acoustical systems so that the signal received on the surface of the mechanical system y (r, t, θ) presents, in the simplest case, the response of the dynamical system with impulse characteristics h (r, t, θ) to the forcing x (t, θ). Here, the space coordinate 'r' is the radius of the vector from the point of reception to the place of stimulation by x (t, θ). Hence, on the basis of the above, and by analogy to Equation (3.2) we can write:

$$y(r,t,\theta) = \sum_{i=1} \varphi_i(t,0) * h(r,t,\theta) * \delta(t - iT) \tag{3.10}$$

(5) The above vibrational process exerts a feedback on the intensity of the wear process and, furthermore, on the technical condition of the machine (subassembly) by a positive destructive feedback (Fig. 3.5). This feedback effect of the destruction process is revealed by a slow change of amplitude and form of the original process φ_i (t, θ).

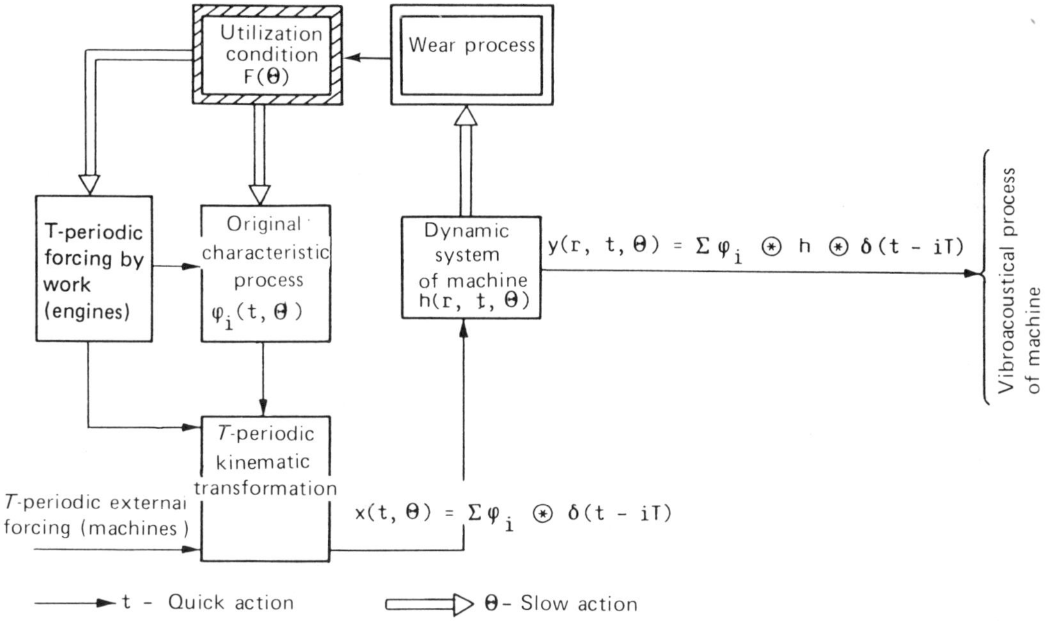

Fig. 3.5 — Cybernetic diagram of the wear model of vibroacoustical phenomena generation in machines [50]

(6) For the determined value of the utilization time θ = const all dynamic systems in our generation model consist of linear and stationary systems whose properties are explicitly described by the impulses response $h\ (r, t, \theta)$ or its Laplace transforms $H\ (r, p, \theta)$, or Fourier transform $H\ (r, j\omega, \theta)$ (the so-called operator and spectral transmittance).

The sequence of assumptions and implications described above is, in fact, a wear (tribological) model of vibroacoustical process generation and it can be presented in the form of a functional model for one failure (Fig. 3.5).

As we know, the source of rotational motion and the motion reactions result in internal forcing as in motors, and in the external forcing, as in the remaining machines and installations. Therefore, it is obvious that the vibroacoustical process $y\ (r, t, \theta)$ must also carry information about the dynamic condition of the internal source of forcing. Such assumption creates the basis for Fig. 3.5 as the general tribological model of vibroacoustical phenomena generations in machines.

3.3.2 Generation of diagnostic signals

Analysing in detail Fig. 3.5, and the sequence of assumptions (1)–(6) leading to it, we can easily arrive at the conclusion that the vibroacoustical process of a machine is a pseudo-T-periodic sequence of events. It will be more obvious, when knowing the influence of the wear feedback, but we omit it for a moment in Fig. 3.5, obtaining in this way Fig. 3.6. The figure facilitates the interpretation of the tribological model of vibroacoustical signals generation since the creation of an observable process $y\ (r, t, \theta)$ can be interpreted in the following way.

The original process $\varphi_i\ (t, \theta)$ is the i-th elementary vibroacoustical event whose form

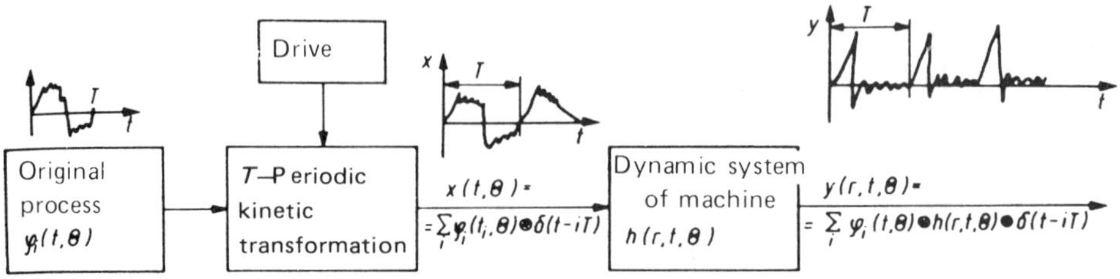

Fig. 3.6 — Transformation of the original process $\psi_i(t, \theta)$ into the vibroacoustical diagnostic process $y(r, t, \theta)$ for θ = const [50]

is determined by the construction and the utilization condition of the machine (subassembly). Owing to T-periodic drive, it it transformed into a sequence of elementary events described as process $x(t, \theta)$. This dynamic process stimulates further vibrations in the whole machine construction giving, in effect, some reception point r, a new transformed sequence of events which is the *vibroacoustical diagnostic process*, i.e. a process used for the identification of the condition of the object.

The presented method of interpretation of the diagnostic signal $y(r, t, \theta)$ is generally true for machines with a periodic action, but it is not always so simple as presented in Fig 3.6. On the contrary, in many cases we have some kind of a complicated system from Fig 3.6. In effect, it gives in one machine a great number of sequences of elementary events $x_i(t, \theta)$, and at the same time they do not have any common period T. The periods of these sequences of elementary events will always be colinear with a certain basic period T_0, i.e. $T_i = a_i T_0$; $0 \leq a \leq \infty$. An example of the necessity of such presentation of the problem is represented by an internal combustion engine whose model of signal generation quoted by the Author [7] is shown in Fig. 3.7.

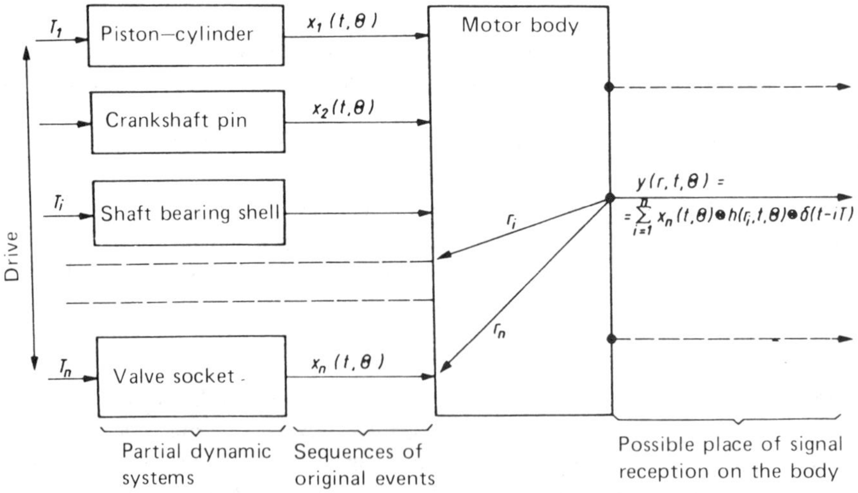

Fig. 3.7 — Detailed model of diagnostic signals generation for an internal combustion engine [7]

It follows from Fig. 3.7 that the diagnostic signal received at any point of the engine structure is the weighted sum of its response to all elementary events x_i (t, θ), and as weights we have here convolutions with impulse transfer functions h (r, t, θ) from the generation place to the reception of the diagnostic signal. So we see that, in many cases, detailed generation models in which the signals can be very complex, carrying a huge amount of information, some of which may not be useful for the purpose of diagnosis. An extreme example would be the vibrations of the body of a jet engine with a multistage turbine and compressor with many auxiliary assemblies compactly built into one unit. Here, identifying characteristic signal (e.g. of the thrust bearing of the turbine) is very difficult.

However, in certain cases of machine diagnostics we will separate the diagnostic signal as a sequence of events with comparatively small deformations due to the passage through the dynamic intermediate systems. It will have to do with the situation of internal combustion engines when we estimate phenomena occurring in the piston–cylinder–valves system. For the estimation of the condition of this functional system we will first select three vibroacoustical processes as diagnostic signals: (a) pulsation of the pressure in the carburettor choke; (b) pulsation of pressure in the crankcase; and (c) pulsation of pressure in the area of the exhaust pipe. In each of these processes, particularly in (a) and (b), the diagnostic signal does not deviate much from the elementary sequence of events and, as such, it carries a great load of selective information about the investigated subsystem. In the first signal, the information refers to the tightness of the cylinder system and the condition of the inlet valves; in the second it refers to the tightness of cylinders, and in the third, to the condition of the whole system, together with the quality of the combustion process. Readers interested in these problems are referred to the Author's monograph [7] and more detailed research [51, 52].

The latter example of the pulsation variation in the internal combustion engine has shown the problem of the selection of the proper vibroacoustics process for diagnosing a definite feature of the machine. Information about the tightness of cylinders is carried here by each of the pulsations (a), (b) and (c) above and one should evaluate objectively the content of information about the property concerned in the definite vibroacoustical signal. This problem will be discussed below.

3.3.3 Selection of the observed vibroacoustical process as the source of diagnostic information

The evaluation of the information content about the concerned machine property in the received vibroacoustical process is possible when we pass from the generation model discussed previously to the *information model* of the observed vibroacoustical process. It is facilitated by the fact that, as follows from the previous chapters, the machine properties W are important in diagnostics, i.e. the failures are also the primary sources of the vibroacoustical process. Therefore, we can venture a quantitative evaluation of the information about process w (t), ($W \Rightarrow w$ (t)) in the observed vibration process d (t) or the noise process z (t). The correct course of procedure in the reconstruction of the information model, with the already known generation model, will be reconstructed on the example of a selection between the reception of vibrations, noise or pulsation for the machine with mechanical fault as shown in Fig. 3.8.

As we see from Fig 3.8, the vibrational signal of failure w (t) is subject only to the transformation by the impulse function of transition h_d (r, t), the disturbance by the motion of the neighbouring kinematic pairs of the machine $n_1(t)$ and the dynamic action

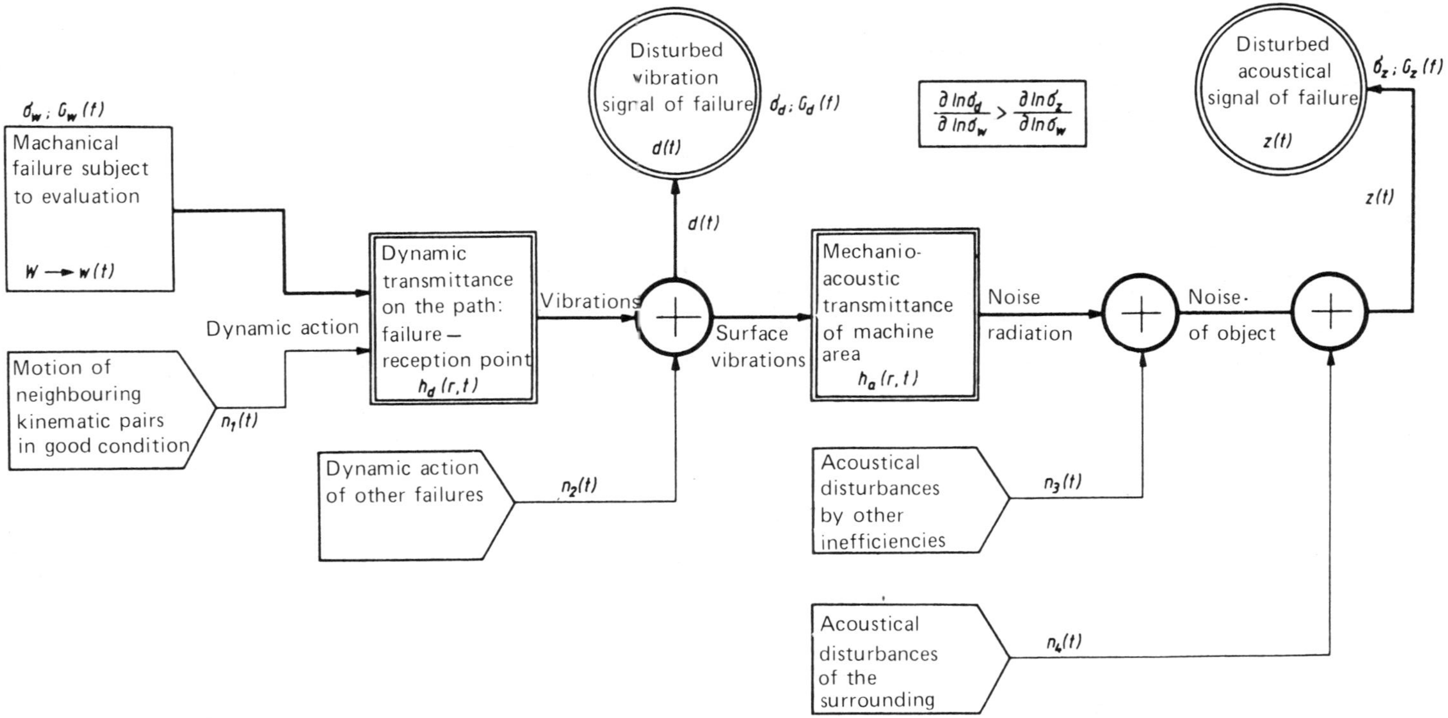

Fig. 3.8 — Information model about the creation of vibrational and noise signals connected with mechanical failure of machine [7]

of other failures $n_2(t)$ (we analyse the situation for θ = const; therefore, we omit this time for simplicity of determination). On the other hand, the acoustical signal $z(t)$ received in the surroundings of the machine carries the already-mentioned disturbances $n_1(t)$, $n_2(t)$ and also the acoustical disturbances of other inefficiencies $n_3(t)$ and the disturbance of the surrounding measurement. Utilizing the laws of the flow of dynamic interactions through the dynamic systems [53] we can write the following relationships for the vibration and noise signals:

$$d(t) = h_d(r,t) * \left[w(t) + n_1(t)\right] + n_2(t) \tag{3.11}$$

$$\begin{aligned} z(t) &= h_a(r,t) * d(t) + n_3(t) + n_4(t) \\ &= h_a(r,t) * \left\{h_d(r,t) * \left[w(t) + n_1(t)\right] + n_2(t)\right\} + n_3(+) + n_4(t) \end{aligned} \tag{3.12}$$

where, as previously, the asterisk denotes the operation of convolution.

The comparison of the notation of the vibration and noise processes speaks for itself and constitutes a further conclusion about the information content following from Fig. 3.8. These conclusions can be extended to the changes of the amplitude of failure signal σ_w when we base on the definition of the sensitivity of signal amplitudes σ_d; σ_z. According to Tomovič [54], the best definition of sensitivity has the form:

$$C \equiv \lim_{\Delta \to 0} \left(\frac{\Delta \sigma_d}{\sigma_d} / \frac{\Delta \sigma_w}{\sigma_w} \right) = \frac{\sigma_w}{\sigma_d} \frac{\partial \sigma_d}{\partial \sigma_w} = \frac{\partial \ln \sigma_d}{\partial \ln \sigma_w} \tag{3.13}$$

and as we can infer from Fig. 3.8 and the dependences (3.11) and (3.12), the following relation is valid:

$$\frac{\partial \ln \sigma_d}{\partial \ln \sigma_w} > \frac{\partial \ln \sigma_z}{\partial \ln \sigma_w} \tag{3.14}$$

where $\sigma(\cdot)$ with an appropriate index denotes the root-mean-square amplitude of the given process $(\cdot)$.

When we want to measure the information content in process $d(t)$ or $z(t)$ about the original process $w(t)$, it is best to take into consideration the quantitative measures of information. The oldest and the most popular measure of Shanon [55] in the simplest model $d(t) = w(t) + n(t)$ is:

$$I = \frac{1}{2} \lg_2 \left[2 + \frac{\sigma_d^2}{\sigma_n^2} \right] \text{ bits}$$

where σ_d^2 is the power of the vibration signal and σ_n^2 is the total power of disturbances in this signal.

It can be used, as we see, only for the evaluation of the capacity of the vibration channel in the presence of noise $n(t)$. Hence, we see that the greater the relation of the signal power to the noise $(\sigma_d/\sigma_n)^2$ the greater the amount of information that can be transferred, e.g. by the vibration signal. This is a very important conclusion indicating the necessity to minimize disturbances in the whole measurement process, and particularly the necessity of selecting an observed process with a minimal content and strength of disturbances.

However, this measure does not permit us to measure the amount of information about the process $w(t)$ in the observed process $d(t)$ or $z(t)$. Introducing the function of coherence as the measure of similarity between the processes $\gamma^2_{dw}(f)$ which we assume to be the stationary one of Gaussian type [56], we can introduce another measure of information:

$$I_{dw} = \frac{1}{2\Delta F} \int_{f\,min}^{f\,max} \lg_2 \left[1 - \gamma^2_{dw}\right]^{-1} df; \quad \text{bits}, \; \Delta F = f_{max} - f_{min} \tag{3.16}$$

The coherence function $\gamma^2_{dw}(f)$ for linear systems is characterized among others by the fact that for the given frequency f, its value decreases the greater the participation of noise $n(t)$ and in process $d(t)$, while in general $0 \le \gamma^2_{dw}(f) \le 1$, unity being the value with a zero noise power [53]. Also, the amount of information transferred in process $d(t)$ about process $w(t)$ will be on the basis of Equation (3.16), the maximal one (f $\gamma^2_{dw}(f) \Rightarrow 1, I_{dw} \Rightarrow \infty$). For example, the coherence function between the vibration process $d(t)$ and the original vibroacoustical process $w(t)$, according to model (3.16) has the form:

$$\gamma^2_{dw}(f) = \frac{\left|G_{dw}(f)\right|^2}{G_{dd}(f)\, G_{ww}(f)}$$

$$= \frac{\left|H_d(f)\right|^2 G^2_{ww}(f)}{G_{ww}(f)\left\{\left|H_d(f)\right|^2 \left[G_{ww}(f) + G_{n_1 n_2}(f)\right] + G_{n_2 n_2}(f)\right\}} \tag{3.17}$$

$$= \frac{\left|H_d(f)\right|^2 G_{ww}(f)}{\left|H_d(f)\right|^2 \left[G_{ww}(f) + G_{n_1 n_1}(f)\right] + G_{n_2 n_2}(f)}$$

where $H_d(f)$ is the vibration transmittance from Fig. 3.8, $G_{ij}(f)$ is the spectral density of the power of processes $i(t), j(t)$, mutual when $G_{ij}(f)$, and own when $G_{ii}(f)$.

We see from Equation (3.17) that the greater the power of the spectral densities in the denominator, the smaller the value of coherence function, and the smaller the amount of information in Equation (3.16) transferred by process $d(t)$ about the source process $w(t)$.

The coherence function, being very convenient in the calculations, has the characteristic that it must be measured synchronically for both processes, which is not always possible. So, let us quote another formula based on the definition of the quantity of *differentiating information*—the Kullback's divergence in the form given by Gersch [57]:

$$I(d,w) = \frac{1}{2} \lg_2 \left(\frac{\sigma^2_d}{\sigma^2_w} \right) + \frac{1}{2\Delta F} \int_{f\,min}^{f\,max} \frac{G_{ww}(f)}{G_{dd}(f)}\, df - \frac{1}{2} \tag{3.18}$$

This way of information content calculation takes into account not only the power of process (σ_d^2 and σ_w^2) but also their own power spectral densities. Therefore, we can calculate from Fig. 3.8 the differences in the range of information between processes *d(t)* and *z(t)*. Calculating by the way also the amount of information transferred in the noise about the source process *w(t)*:

$$I(z,w) = \frac{1}{2}\lg_2\left(\frac{\sigma_z^2}{\sigma_w^2}\right) + \frac{1}{2\Delta F}\int_{f\,\min}^{f\,\max}\frac{G_{ww}(f)}{G_{zz}(f)}\,df - \frac{1}{2} \tag{3.19}$$

and creating differences of Equation (3.19) we have:

$$\Delta I(d,z) = I(d,w) - I(z,w)$$
$$= \frac{1}{2}\lg_2\left(\frac{\sigma_d^2}{\sigma_z^2}\right) + \frac{1}{2\Delta F}\int_{f\,\min}^{f\,\max}\left[\frac{G_{ww}(f)}{G_{dd}(f)} - \frac{G_{ww}(f)}{G_{zz}(f)}\right]df \tag{3.20}$$

Owing to a greater content of disturbing noise in the noise process *z(t)* which follows from the comparison of Equations (3.11) and (3.12), we can assume

$\frac{G_{ww}(f)}{G_{dd}(f)} > \frac{G_{ww}(f)}{G_{zz}(f)}$. Thus, from Equation (3.20),

even with equal powers of vibrations and noise ($\sigma_d^2 = \sigma_z^2$), we have $I(d, z) > 0$ which affirms the informational superiority of vibrations over noise in the diagnostic situation analysed in Fig. 3.8.

An opposite situation to the previously analysed one can be met in failures which give effects of a flow-through type as, for example, in the evaluation of the tightness (or leakage) of a piston–cylinder in an internal combustion engine (Fig. 3.9). From the data in Fig. 3.9 we can draw the conclusion that the sensitivity of pulsation *z(t)* is greater for the diagnosed fault *w*, since it is easy to show the opposite property than that in Fig. 3.8, i.e.:

$$\frac{\partial \ln \sigma_z}{\partial \ln \sigma_w} > \frac{\partial \ln \sigma_d}{\partial \ln \sigma_w} \quad \text{and} \quad \Delta I(z,d) > 0 \tag{3.21}$$

These two examples of the selection of diagnostic signals suggest the conclusion that, having information models of vibroacoustical processes, we can select each time a process with a greater sensitivity to changes of the investigated property *W*. This sensitivity (i.e. the amount of information carried about property *W*) can also be evaluated in a quantitative way according to the given relations when we have data about the spectra of the analysed processes and of the mutual processes. Even without these data we can draw an important application conclusion from Equation (3.20) if processes *d(t)* and *z(t)* are equally sensitive to changes *w(t)*, and when they differ only by the amplitude, since from Equation (3.20) we have:

$$\Delta I(d,z) = \frac{1}{2}\lg_2\left(\frac{\sigma_d}{\sigma_z}\right)^2 \tag{3.22}$$

when $\frac{G_{ww}(f)}{G_{dd}(f)} \equiv \frac{G_{ww}(f)}{G_{zz}(f)}$.

It indicates that, with equal spectral sensitivity, we must apply a process with a greater

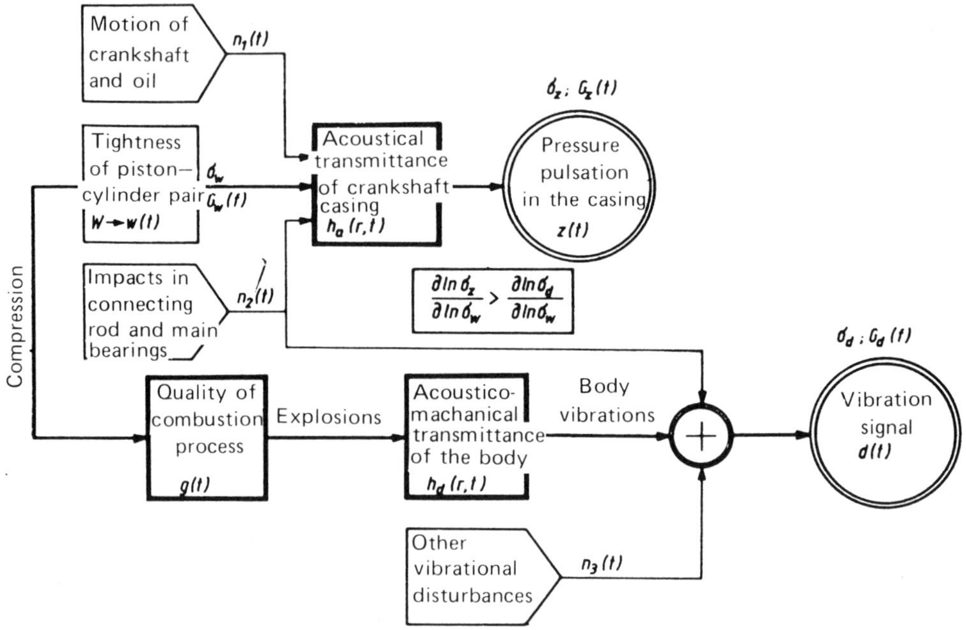

Fig. 3.9 — Information model of the creation of vibration and pressure pulsation signals in the evaluation of the tightness of the piston–cylinder in and internal combustion engine

power which gives $\Delta I > 0$. Hence, we can write an important statement for vibrational diagnostics: *Measure vibrations in the direction of the greatest amplitude.* Since the rigidity of many machine nodes (e.g. the housing of the bearings) is smaller in the horizontal direction, we have here greater amplitudes of vibrations and therefore we make a greater number of diagnostic measurements in this direction. And the place of vibrational measurement should be selected *as close as possible to the potential* failure, where the stream of vibration energy is maximal.

3.4 SELECTION OF VIBROACOUSTIC SIGNALS

We know which vibroacoustical processes to choose for diagnostic purposes, what direction of observations to apply and how to evaluate the information content of one process in another one. In this situation it would seem that it is all that we need for vibroacoustical diagnostics purposes. However, when we recall the general expression in Equation (3.2) for the observed vibroacoustical process:

$$s(r,t,\theta) = \sum_{i=1}^{n} h(r_i, t, \theta) * V_i(t,\theta) + c(r,t,\theta) + n(r,t,\theta)$$

and we realize that our diagnostic objective requires the valuation of the original vibroacoustic process (i.e. $V_k(t, \theta)$) then we will quickly stop being enthusiastic. Namely, there emerges immediately the problem formulated in the title and consisting in the preliminary selection of the required signal, i.e. in the selection of such an operation which will give at least an approximation of:

$$WS_k\{s(r,t,\theta)\} \cong h(r_k,t,\theta) * V_k(t,\theta) = s_k(r,t,\theta) \qquad (3.23)$$

where $WS_k\{\cdot\}$ is the symbolic operator of the preliminary selection of the k-th original signal of failure.

Looking at the relation in Equation (3.23) we come to the conclusion that this operation can be carried out in two fields r and t in which the instantaneous values of the vibroacoustical process is decided. And when we take into consideration the spectral presentation of the vibroacoustical processes by Fourier's transformation (see, for example, Bendat and Piersol [53]) then we will obtain an exchangeability of the field of time 't' into the field of frequency 'f'. In some cases of selection, particularly in the utilization of wave motion properties, we observe also the exchangeability of the field of dynamic time t' with the coordinate of place r. Since there is a connection through the speed of sound c_0, the distance covered along one coordinate can be expressed as $r = c_0 t$, or sometimes as the difference of paths $\Delta r = c_0 \Delta t$. This will be developed, concentrating on a preliminary selection of diagnostic signals of our hypothetical k-th failure and at the same time, next to the three fields of signal determination 'r, t, f' already mentioned, we will also utilize the information about the temporal character of the signal itself $V_k(t, \theta)$.

3.4.1 Spatial selection

The spatial selection consists of utilizing the properties of the wave field of the received vibroacoustical phenomenon and the properties of the system of transducers or special reception antennae. It is best explained by a hypothetical example similar to reality: Let noise be the observed phenomenon generated by a group of spinning spindles in which we investigate the condition of the rolling element bearings. The progressive failure of a bearing gives an increased level of high-frequency vibrations (15kHz and more) which is emitted as noise. With the availability of a selective directional microphone, we can carry out a survey of the actual condition of bearings by turning the microphone to the corresponding directions, in Fig. 3.10. For higher frequencies (e.g. $f > 15$ kHz), a directional microphone can be easily made by putting on a simple microphone an exponential horn with index β. Then, for the frequencies

$$f > 2.3 f_k, \left(f_k = \frac{c_0 \beta}{4\pi} \right)$$

[58], we can obtain a transducer with sufficient directional characteristics for distinguishing signals of the particular spindles.

The horn here is the simplest example of an antenna system reacting to differences in the transition time of the wave front to the different places of horn input. More advances antennae are used as, for example, transducer systems in geophysics, in location of earthquakes and underground explosions, in underwater acoustics for the location of ships, and in acoustical emission for the location of cracks, etc. Here, the general principle is that the greater the quotient of the antenna dimension D in relation to the length of the received wave λ, $(D/\lambda \gg 1)$ the better its directional selectivity.

If the dimensions of the phenomenon and the reception field D are small in relation to the wave length λ, as in vibrations of elastic bodies with the frequency of the order

10–100 Hz, then the antennae utilizing the difference of transition times are useless, since the phenomenon in the whole measurement field takes place almost simultaneously. The reception antennae (systems of transducers) must be sensitive to the direction of motion. Vibration transducers have rather strong directional properties so, when we position them adequately and sum up the signals received from them, we obtain an elimination of some signals (phenomena) and the amplification of others.

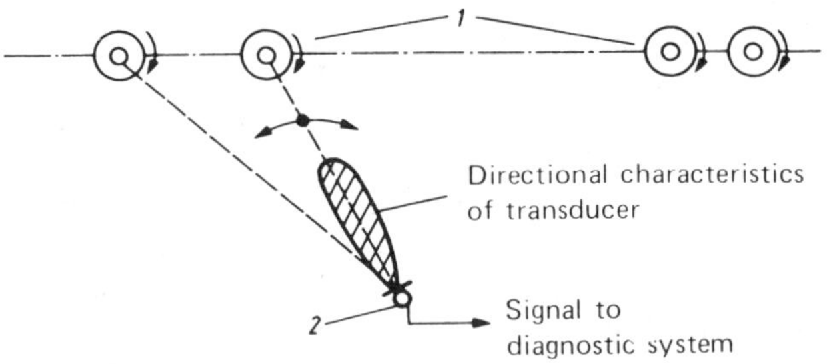

Fig. 3.10 — Spatial selection of the signals of the bearings of spinning spindles with the help of a microphone with strong directional characteristics

An example of such reasoning is given in Fig. 3.11, where we see that, with the help of two vibration transducers fixed on two opposite walls of a cylinder, the vibration signal of the whole block can be eliminated; on the other hand, with the help of four transducers we can receive a pure signal of the reversible strokes of the piston against the cylinder.

From the examples quoted above we see that by utilizing the properties of the wave motion, or by using our previous knowledge about the generation of low-frequency vibrations, we are able to design a system of vibroacoustical transducers preliminarily separating the useful signal of the investigated failure from the total vibroacoustical signals of the machine.

3.4.2 Time selection of vibroacoustical signals

The time selection is possible when we know the time sequence of the investigated events in relation to a definite reference signal. In this case, the reference signal controls the gate of of the electronic strobing system opening the gate after strictly defined time periods T, and with opening intervals $\Delta \tau$. At the same time, it is obvious that these times follow from the construction and the functional structure of the machine. Owing to its nature, the time selection is applied to the filtration of impulse signals, impact signals in kinematic pairs, and mechanical signals where the sequence of phenomena is essential, e.g. in the opening/closing of the valve, and the striking of the ram on the tool in a pneumatic hammer, etc. A hypothetical example of the action of such a system of preliminary time selection of diagnostic signals is presented in Fig. 3.12. As we see, after the selection of the useful signal, we can evaluate each fragment of it in three different ways: (a) time of duration Δt_k; (b) amplitude s_k; and (c) deviation from the expected synchronism e_k. The time selection can also utilize the time differences of wave arrival in connection with the spatial localization of the presumed failure. Then we have to consider the spatial and time localization, this having been already used for the supervision of the integrity of definite

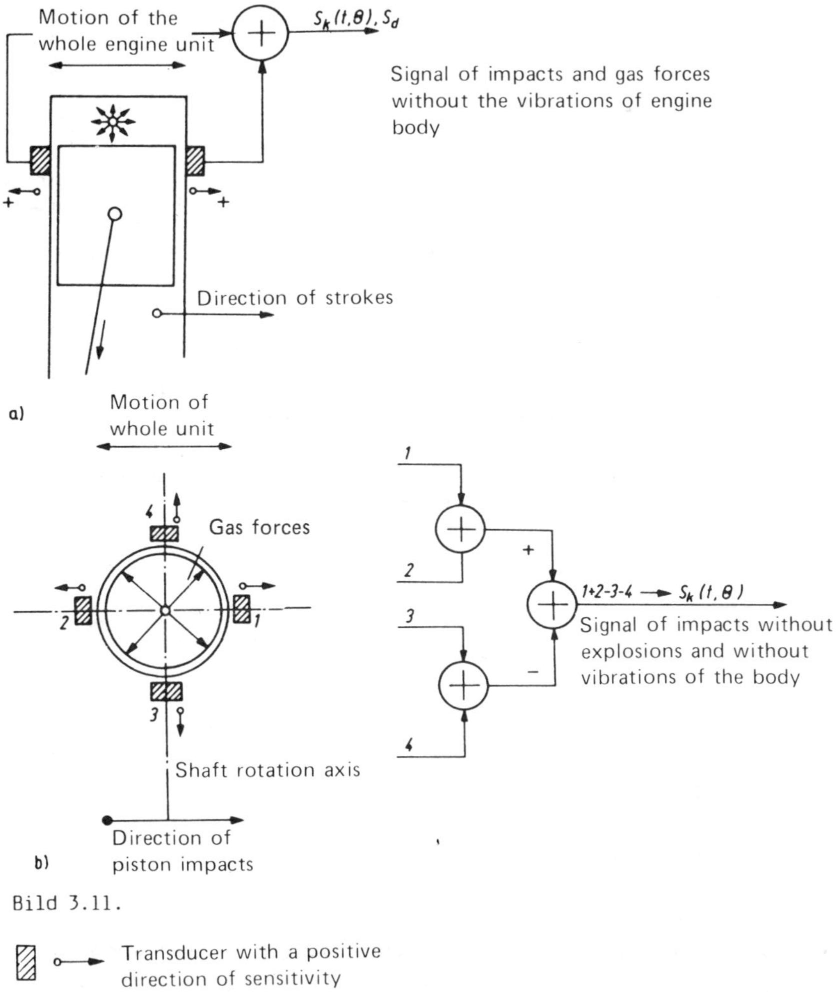

Fig. 3.11 — Examples of the localization and summing up of the vibrations transducer signals for the selection of the strokes of the piston against the cylinder [7]

construction places with the help of acoustical emission. A typical example of the application of spatial and time selection based on the coincidence effect of signal transmission from a definite zone is the supervision of cracks around the assembly openings of aircraft wings [59]. The coincidental nature of the selection of signals is shown in Fig. 3.13, in which the signal of canal A opens the gate to interval $\Delta\tau$ following from the width of a supervised zone. The accepted signals must be contained in this time interval $\Delta\tau$, as in the case of strobing (Fig. 3.12).

3.4.3 Spectral selection

In vibroacoustical diagnostics, we frequently do not know the sequence of events in order to be able to apply time selection, but we know the form of the event which is to happen.

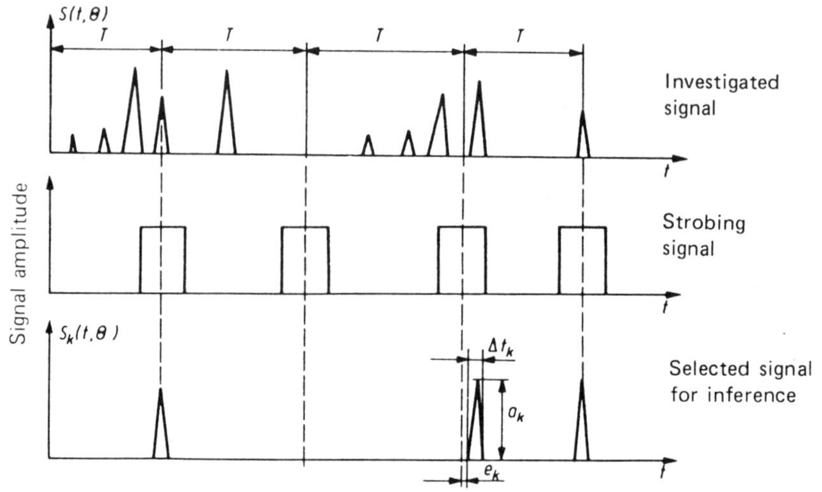

Fig. 3.12 — Example of time selection of a complex diagnostic signal with a known reference signal

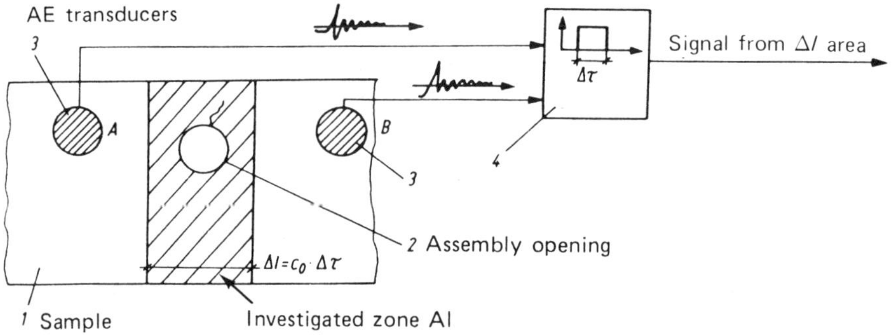

Fig. 3.13 — Time and spatial selection of acoustical emission signal with the help of transducers and the coincidence system [59

From the shift theorem of transform [60] it follows that the time shift of the phenomenon (unknown) does not change its modulus of Fourier's spectrum since:

$$\left|\Im\left[V_k(t-a)\right]\right| = \left|e^{-2\pi ifa}\,\Im\left[V_k(t)\right]\right| = \left|\Im\left[V_k(t)\right]\right| = \left|V_k(f)\right| \qquad (3.24)$$

If $h(t)$ is an impulse transfer function of a system (electronic, electric, pneumo-hydro-mechano-acoustic system), then its transmittance is $H(f)$, while the spectrum of the delayed (accelerated) signal by 'a' is the modulus equal to the spectrum of the original signal. So, if we know at least an approximated form of the original signal of failure, then by applying a filter with a similar or identical transmittance $H(f) \simeq V_k(f)$ we will obtain the convolution of signal $s(t, \theta)$ with the impulse transfer function of transition, or in the field of spectrum we will get their products as in Equation (3.25) below:

$$
\begin{aligned}
&S(F,\theta)V_k(f) \\
&= \left[\sum_{i=1}^{n} H\left(r_i, f, \theta\right) V_i(f,\theta) + C(r,f,\theta) + N(r,f,\theta) \right] \cdot V_k^*(f) \qquad (3.25) \\
&\cong H\left(r_k, f, \theta\right) \left| V_k(f,\theta) \right|^2 + N_k(r,f,\theta)
\end{aligned}
$$

As we see, this operation is equal to the minimization of all signals and noises and the squaring of the spectrum of the investigated tribovibroacoustical process which is a very advantageous operation. Even if we do not know the form of the searched-for tribovibroacoustical process, but we know its timescale (e,g quick/slow or high-/low-frequency) then in order to decrease the relation of the signal to noise we apply the low-pass, medium-pass or high-pass band filtration with the transmittance function of the rectangular window type. The fact that such an approach gives essential diagnostic advantages is easy to see when analysing Fig. 3.14 quoted after Balderstone [61].

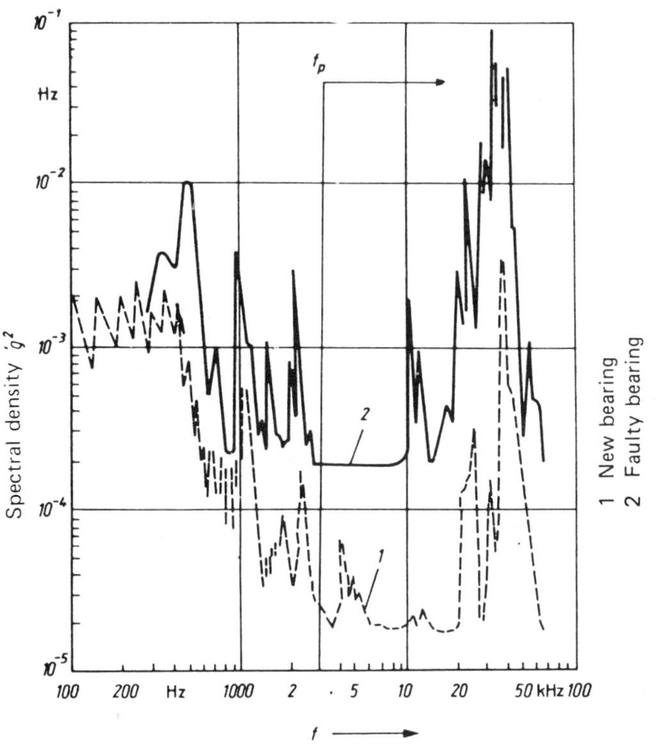

Fig. 3.14 — Spectral density of the acceleration of two ball bearings [61] with the indication of the possibility of distinguishing the diagnostic band through a high-pass filter $f_d > 3$ kHz

The figure shows distinctly that the high-pass filtration of this definite signal gives unmistakable differentiation of the quality of the rolling bearing.

Referring to general properties of the spectra of tribovibroacoustical processes, it is worth mentioning that the scale of the phenomenon has its reflection in the theorem about the similarity of Fourier's spectra [60], where:

$$\Im\left[V_k(a, t, \theta)\right] = \frac{1}{a} V_k\left(\frac{f}{a}, \theta\right) \tag{3.26}$$

and depending on factor 'a' the spectrum of the process can be a low-frequency $a \gg i$, medium-frequency $a \sim 1$, and upper-frequency one $a \ll 1$ which, as has already been shown, is utilized in the preliminary band filtration of tribovibroacoustical signals (Fig. 3.14).

The majority of machines and installations, performing periodic work stationary motions. Therefore the valid generational model is presented in Section 3.3.1 (Fig. 3.5 and Equation (3.10)); hence it follows that the generated tribovibroacoustical process is also a periodic one. Therefore, each disturbance of this periodicity can give information about the technical condition of the object; therefore, it is necessary to select the periodic part of the tribovibroacoustical signal. Analytically it can be easily done by giving to the filter transmittance $H\,(f)$ the form of as delta sequence in the field of frequency (i.e. a comb filter) as below:

$$H_T(f) = \frac{1}{T_0} \sum_{i=0} \delta(f - if_0), \quad f_0 = \frac{1}{T_0} \tag{3.27}$$

where T_0 is the basic period of the machine work.

When we multiply the spectrum of our general signal $S\,(f, \theta)$ by the assumed transmittance as the effect of filtration, i.e convolution in the domain of time, we have:

$$\begin{aligned} S\,(f, \theta)\, \mathrm{H}_T(f) &= S\,(f, \theta) \frac{1}{Z_0} \sum_{i=0} \delta(f - if_0) \\ &= \frac{1}{T_0} \sum_{i=0} S_T(if_0, \theta)\, \delta(f - if_0) \\ &= \Im\left[s(t, \theta) * \sum_{i=0} \frac{1}{T_0} \delta(t - i\,T_0) \right] \end{aligned} \tag{3.28}$$

where $S_T\,(i\,f_0, \theta)$ is the periodic part of the signal from $S(f, \theta)$ with successive harmonics $i = 1, \ldots,.$

As we see, with the help of an infinite comb filter we have obtained a perfect separation of the periodic component of the diagnostic signal. In practice, it is not so simple, owing to the finite number of filter members (see, for details, the Author's description of summing up and synchronic analysis [7]).

If we look at Equation (3.28) knowing the properties of Fourier's transform [60, 62], we will notice that it is a picture of the periodical process consisting of T_0 periodic repetition of the original process:

$$\begin{aligned} &\frac{1}{T_0} \sum_{i=0} S_T(if_0)\, \delta(f - if_0) \\ &= \Im\left[\sum_{1=0} s_T(t, \theta) * \delta(t - iT_0) \right] \\ &= \Im\left[\sum_{1=0} s_T(t - iT, 0)\, u(t - iT) \right] \end{aligned} \tag{3.29}$$

where $\Im[\cdot]$ is Fourier's transform, the asterisk is the convolution in the field of time and $u(q)$ is the Heaviside function similar to the tribovibroacoustical model of signal generation in Equation (3.9).

Comparing this result with the generation model of signals in Equation (3.9), we see that the periodically repeated signal $s_T(t, \theta)$ is an analogue of the original tribovibroacoustical process $\psi_0(t, \theta)$ as in Equation (3.8).

The operation in Equation (3.29) carried out in the field of time is called synchronous summation and owing to the infinite number of members, it has no noise component containing only a non-deformed signal. In fact, the number of summed fragments is always finite involving the operation error (see the Author's findings for details [7]).

Summing up the spectral possibilities of the selection of tribovibroacoustical signals let us take a look at Fig. 3.15, which gives a compact representation of the idea of the four discussed methods of tribovibroacoustical signal filtration.

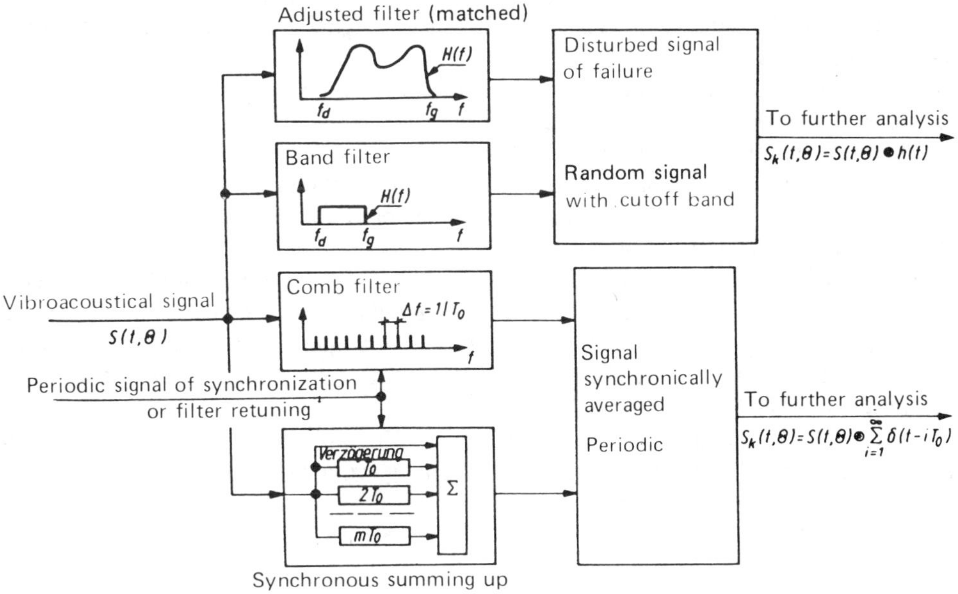

Fig. 3.15 — Alternatives of the matched, band and periodical spectral filtration of vibroacoustical signal

3.4.4 Selection of vibroacoustical signal by demodulation

In many cases vibroacoustical diagnostics undertaken in the presence of the background of a vibrational high-frequency (kilohertz) signal which does not carry any information about the machine condition, here low-frequency components are revealed carrying information about the forcing and the advancement of tribovibroacoustic processes in the mechanism or in the machine. We conclude, then that the high-frequency process is a carrier component and the low-frequency process is the modulation component. Examples of such signals can be found in almost every machine, but most frequently they are revealed in rolling element bearings and transmission gears. In both cases the frequencies and the carrying processes are the resonances of such elements as bearing raceways, toothed wheel rims and gear housings, etc. The modulating phenomena are the impacts of elements (e.g. in bearings, gears and clutches, etc.) which give, in general, a modulation of signal amplitude. On the other hand, the variable stiffness of bearings, tooth and shaft, etc. as a function of rotation angle, enables such elements to give the frequency modulation of the carrying signal. Generally, we can say that the reasons for parametric vibrations in the machine (e.g. the ones mentioned above) are at the same time the causes of frequency modulation of the generated vibroacoustical signal (Sections 2.2.3 and 2.2.6).

In a general case of high-frequency vibrations of machines, both types of modulation take place simultaneously and at the same time, depending on the definite case of a mechanism, we have the superiority of amplitude modulation or frequency modulation. Such phenomena of mixed modulation can be represented by a simple mathematical notation:

$$s(t) = A(t)\cos\left[\omega_n t + \psi(t)\right], \; \omega_n = 2\pi f_n = \text{const} \tag{3.30}$$

where A (t) is the slowly variable process modulating the amplitude of the phenomenon, $\psi(t)$ is the quickly variable process modulating the signal phase and f_n is the carrier frequency of the signal.

Since phase modulation is equal to frequency modulation we will only consider two types of modulation. So, from the latter formula for $\psi(t)$ = const, we will have a pure modulation of amplitude by slowly variable process A (t) = var; on the other hand, if A $(t) \equiv$ const ψ (t) = var, we have a pure frequency modulation through a quickly variable process ψ (t), which changes the resultant frequency of vibrations around the basic carrier frequency ω_n such as in the parametric vibrations.

In order to extract the diagnostic information from the process transformed in such a way by modulation, we use different types of demodulators. A classical example of such possibilities in everyday life is the simple radio set working in long, medium and short waves as the demodulator of amplitude (AM) and a demodulator of frequencies for ultrashort waves (FM).

The amplitude demodulation is nothing but the utilization of the envelope of vibroacoustical processes which can be descriptively compared to the envelope of the maxima of this process, as in Fig. 3.16.

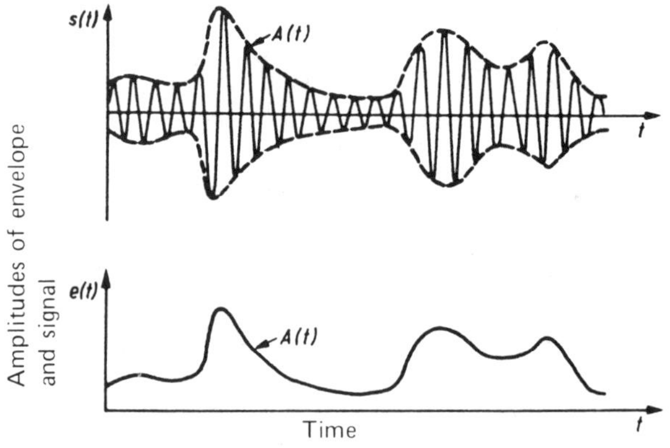

Fig. 3.16 — A vibroacoustical signal and its envelope e (t) after demodulation of amplitude

The simplest analytical notation of the amplitude demodulation operation can be presented as the current averaging of the successive maxima of the original signal:

$$e(t) = \underset{\tau}{E}\left[\max_{t} |s(t)|\right] = \frac{1}{\tau}\int_{t}^{t+\tau} \max \left|A(t) \sin\left[\omega_0 t + \psi(t)\right]\right| dt \qquad (3.31)$$

If the phase of signal is constant ($\psi(t)$ = const) then from this operation with the optimally selected time of current averaging $\tau \gg \frac{2\pi}{\omega_0}$ we will obtain :

$$e(t) = \frac{1}{\tau}\int_{t}^{t+\tau} A(t)\, dt \sim A(t) \qquad (3.32)$$

From the above we can infer that the demodulation of amplitude will be better, the narrower the band in the original vibroacoustical process s (t). Hence, frequently the demodulation is connected with a preliminary filtration giving good results. This conception is illustrated in Fig. 3.17, by a precursor of diagnostics, Pavlov [63, p. 162]. It follows clearly from the Figure how the filtration, in this case the high-pass filtration, visualizes the real depth of the modulation of the vibroacoustical process and increases thereby the amplitude of signal after demodulation.

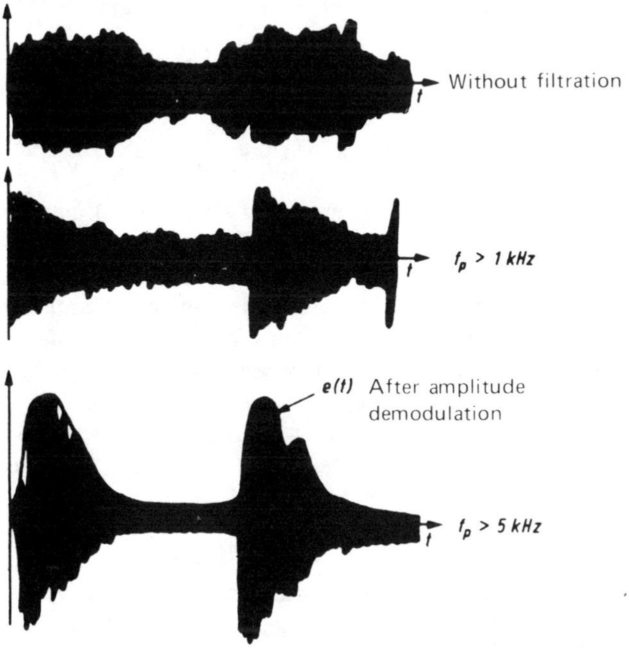

Fig. 3.17 — Preliminary selection of vibroacoustical signal by the high-pass filtration and demodulation of amplitude

The more recent possibility of amplitude demodulation as the modulus of analytic signal being the combined sum of the signal and its Hilbert's transform [62, p. 324] is significant and in the age of computerization is a very convenient tool for amplitude demodulation, but we are not going to deal with it any further here.

Let us take into consideration the frequency demodulation $A(t) = \text{const}$, $\psi(t) = \text{var}$. Here, in order to extract information from the signal illustrated in Equation (3.30) and in a general cause from the band signal with medium frequency f_n (e.g. of the octave band with the medium f_n), a signal proportional to the instantaneous frequency difference should be worked out so that:

$$e(t) = \omega_n - \frac{d\psi(t)}{dt} \tag{3.33}$$

For better clarity of this conception, let us carry out the following reasoning. For the harmonic process $x(t) = A \cos \omega_n t$ we have the following connections between the amplitudes of the initial process and its derivative:

$$\text{Ampl}\,\frac{dx}{dt} = |\dot{x}(t)| = \omega_n \text{ ampl } x(t) \tag{3.34}$$

Hence, by analogy we can write that the instantaneous frequency of modulation $\Omega(t) = [d\psi(t)]/dt$ will be:

$$\Omega(t) = \frac{d\psi}{dt} = \frac{\text{ampl}\,\frac{ds(t)}{dt}}{\text{ampl } s(t)} = \frac{\text{ampl } \dot{s}(t)}{\text{ampl } s(t)} \tag{3.35}$$

In such a situation the frequency demodulation can be presented as a difference signal from the carrier ω_n

$$e(t) = \omega_n - \Omega(t) = \omega_n - \frac{\text{ampl } \dot{s}(t)}{\text{ampl } s(t)} \tag{3.36}$$

which is no problem when we can generate the quotient of amplitudes of the derivative and the original signal according to Equation (3.35).

To conclude our discussion of demodulation as a technique of preliminary selection of a vibroacoustical signal, consider Fig. 3.18 which presents the application of both techniques for the signal of housing vibrations of a high-speed industrial gear [64]. As we see, the demodulation of the frequency around the frequency meshing f_x and the demodulation of amplitude reveal the whole richness of harmonic rotations of low- and high-speed rollers which is not visible in the acceleration spectrum. On the other hand, the spectrum of the signal after each demodulation is more readable than the spectrum of vibration velocity of the gear housing.

3.4.5 Decomposition of signals and periodic adaptational filtration

Section 3.4.3 was devoted to the spectral selection, whenever we have shown that in some cases it is useful to make a spectral decomposition of the signal into components in respect to some bands. It will be even more useful to distinguish the periodic components if we know a priori the sequence and period T_0 of events, or the diagnostic band of our tribovibroacoustical signal. Such division or rather decomposition of tribovibroacoustical signal may be done automatically with the help of some installations or computer algorithms after specifications of decomposition criteria. Taking into consideration the most general model of tribovibroacoustical signal in the form of Equation (3.2) we have in the known notation:

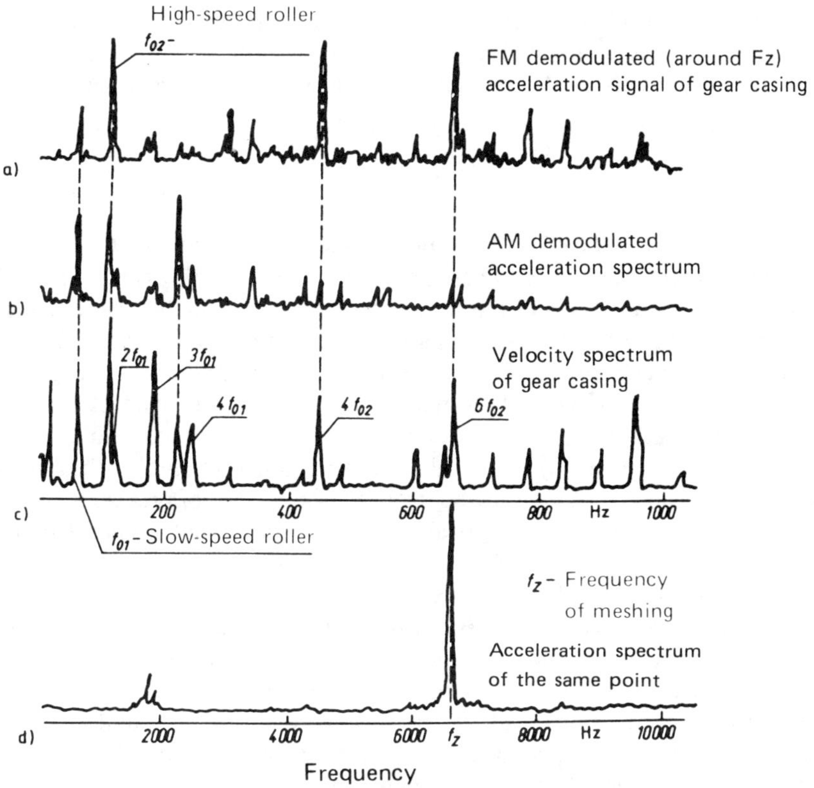

Fig. 3.18 — A comparison of the spectra of gear housing vibrations with the utilization of the techniques of frequency and amplitude demodulation [64]

$$s(r,t,\theta) = \sum_{i=1}^{n} h_i(r,t,\theta) * V_i(t,\theta) + c(r,t,\theta) + n(r,t,\theta) \tag{3.37}$$

The resulting ideal system (algorithm) of the decomposition of this signal is presented in Fig. 3.19.

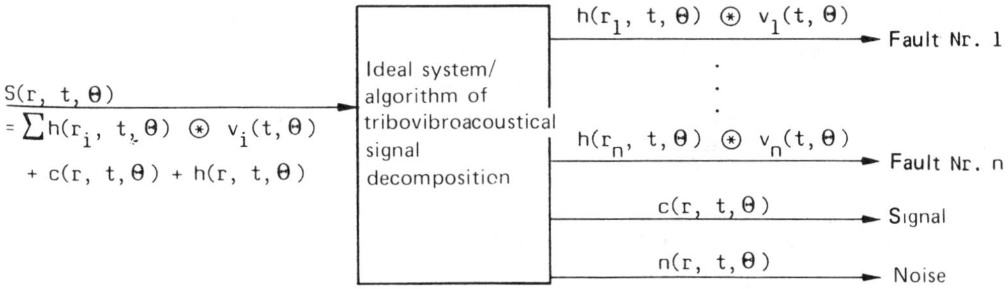

Fig. 3.19 — Concept of an ideal decomposition of tribovibroacoustical signal

When we are able to specify the common and differentiating features of the particular components, we will be able to think about the principle of decomposition. The first thought is the possible assumption that there is no mutual correlation between them, i.e.:

$$E_t\left[v_i v_j\right] \equiv 0; \quad E_t\left[n v_j\right] \equiv 0;$$
$$E_t\left[c\,n\right] \equiv 0; \quad E_t\left[c v_j\right] \equiv 0; \qquad i, j = 1, \ldots, n \tag{3.38}$$

where $E_t\,[\cdot]$ denotes the operator of mean value in respect to time t.

The first two assumptions about the independence of the original tribovibroacoustical processes of each other and the noise are almost obvious but the independence of the rated working process c (r, t, θ) can evoke objections in reference to some machines. However, if we assume that this is true, then referring to our knowledge from the theory of signals and the theory of estimation we can present at least three algorithms of decomposition according to Fig. 3.19, which can lead to similar results.

Deutsch [65, p. 64] gives an algorithm of Gram-Schmidt orthogonalization in the simplified version by Cameron who, from an optional set of processes (system of random vectors), creates a set of orthogonal processes, e.g. of our original tribovibroacoustical processes v_i (t, θ). The problem consists here in the fact that we do not have at our disposal the initial set. However,this can be easily overcome by creating for this purpose from the original process s (r, t, θ) the differential processes $s^{(m)}$ (r, t, θ), and the integral processes $s^{(-m)}$(r, t, θ) and then carrying out the decomposition (here the power index ($-m$) denotes the order of integration and (m) the order of differentiation).

The second possibility with the same initial data can be found in the conception of the partial coherence function of a higher order proposed by Dodds and Robson [66]. However, this problem has not been investigated.

The third possibility of decomposition lies in the method by Karhunen-Loeve [64, Chapter 5] where the decomposed partial processes are found as the solution of the eigenproblem of signal s (r, t, θ) correlation function.

Having outlined these possibilities, let us leave them aside until Chapter 6 where we will proceed to advanced problems. Here we want only to call attention to the common feature of these decomposition techniques. Their initial data (e.g. V_i (t, θ)) change with the change of input s (r, t, θ). Therefore, the mentioned procedures of decomposition must have an adaptational character, conversely than in the decomposition of the process into the components of Fourier's spectra with 'rigidly' determined basic functions sin $2\pi f_n t$ and cos $2\pi f_n t$.

On the borderline of these two opposite possibilities (i.e. the 'rigid' harmonics of basic functions and the basic functions deducted from the analysed process) we can place the periodic adaptation filtration. The periodic filtration has been discussed and described with the help of the transmittance of an ideal comb filter in Equation (3.26), but we still need the signal for an automatic tuning of the filter to the period of the analysed signal. It can be done by investigating the current correlation of the original and the differential signals after filtration as in Fig. 3.20.

As shown in Fig. 3.20, the input signal is T-periodic and the filter is preliminarily set as T_0-periodic. Taking the differences of signals at the output we have:

$$x(t) = s(t) - s_{T_0}(t) = s_T(t) + n(t) - s_{T_0}(t) \tag{3,39}$$

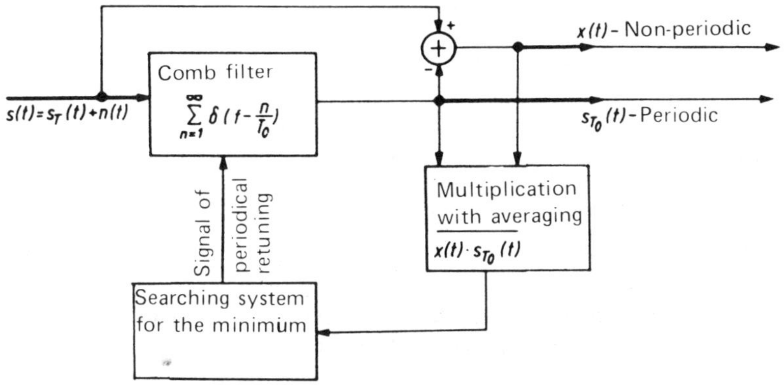

Fig. 3.20 — Concept of an adaptive periodical filter for the initial separation of the signal into periodic and non-periodic parts

and calculating the correlation of output signals for the time of delay $\tau = 0$ we obtain:

$$R_{xs_{T_0}}(0) = E_t\left[x(t)\, s_{T_0}(t)\right]$$

$$= E_{t_0}\left\{\left[s_T(t) + n(t) - s_{T_0}(t)\right] s_{T_0}(t)\right\}$$

$$= E_t\left[s_T(t)\, s_{T_0}(t)\right] + E_t\left[n(t)\, s_{T_0}(t)\right]^{0} - \overline{S^2_{T_0}(t)}$$

Taking into consideration the properties of correlation function [53, Chapter 3] and the absence of noise correlation $n(t)$ in each of the processes, we have the first and the last members as non-zero members. They are mutually reduced if the minimum searching system tunes the periodicity of filter T_0 to the T-periodicity of the signal, since:

$$\lim_{T_0 \to T} R_{xs_{T_0}}(0) = \left\{E_t \left| s_T(t)\, s_{T_0}(t) - \overline{S^2_{T_0}(t)}\right.\right\}_{T = T_0} = 0 \qquad (3.40)$$

So, we see that the periodical filter from Fig. 3.20 can be adapted to an unknown signal period at the input and thereby can decompose the signal into the T-periodical and the non-T-periodical components. In many cases the input signal can be multiperiodical, that having already been mentioned [7] in proposing a special synchronical analysis. However, we are not going to investigate these questions any further here, but will return to the advanced problems of vibroacoustical diagnostics in Chapter 6, notably for further adaptive possibilities of signal filtration by numerical adaptation filters (the one described above can have also an analogue realization).

Concluding the considerations about the preliminary selection of vibroacoustical signals, we will show its need in the categories of the amount of information. Coming back to Shanon's [55] formula given already by the Equation (3.15) let us write it for a general type of tribovibroacoustical signal (Equation (3.36)) with the assumption that we are looking for the original k-th signal of failure $h(r_k, t, \theta) * V_k(t, \theta)$. Adapting Equation (3.15) to these conditions we will find:

$$I = \frac{1}{2}\lg_2\left[1 + \frac{\overline{\left[h(r_k, t, \theta) * V_k(t, \theta)\right]^2}}{\overline{\left[s(r, t, \theta) - h(r_k, t, \theta) * V_k(t, \theta)\right]^2}}\right]$$

$$\cong \frac{1}{2}\lg_2\left[1 + \frac{\overline{\left[h(r_k, t, \theta) * V_k(t, \theta)\right]^2}}{\overline{[s(r, t, \theta)]^2} - 2\overline{s(r, t, \theta)\, h(r_k, t, \theta) * V_k(t, \theta)} + \overline{\left[h(r_k, t, \theta) * V_k(t, \theta)\right]^2}}\right] \quad (3.41)$$

where the line over the brackets is used here exchangeably with the operator of averaging $E_t[\cdot]$.

Let us look at the denominator of the latter expression and let us realize that the signal selection is, in fact, a limitation of the number of tribovibroacoustical components:

$$s(r, t, \theta) = \sum_{i=1}^{n} h(r_i, t, \theta) * V_i(t, \theta) + c(r, t, \theta) + r(t, \theta)$$

In an extreme case of ideal selection of the above signal oriented to the k-th failure, we will obtain zero in the denominator Equation (4.41), i.e. an infinite ability of transferring information about the investigated failure. So we see that the application of the proper method and instrument to the preliminary selection of vibroacoustical signals minimizing the number of superfluous components is part of the diagnostic success.

In conclusion, in the presentation of signal selection methods we can state that it is an important tool for increasing the signal to noise ratio. The more so, because many methods can be applied here in combination. And so, the spatial selection can be almost always applied in connection with the spectral selection. Even a threefold selection of vibroacoustical signal is possible, e.g. after the spatial selection, the cutting off of spectrum and amplitude demodulation, or phase demodulation depending on the needs.

3.5 DIAGNOSTIC MEASURES OF VIBROACOUSTICAL SIGNALS AND THEIR APPLICATION

3.5.1 Introduction

We have already considered in this chapter the possible models of the generation of tribovibroacoustical phenomena, and we have premises for the selection of the diagnostic process, the point and direction of measurement. We also know how to select preliminarily the useful diagnostic signal from the whole wealth of the possible information flowing in the observed diagnostic process. Therefore, we need now to apply to the selected diagnostic process the different types of measures—the simple, complex, and special ones—and we must clarify their application and their diagnostic possibilities. We will start this presentation from *simple estimators* namely, root-mean square amplitudes or peak amplitudes of vibrations and noise, being called the *point measures* (numerical measures). Then we will generalize the notion of a point estimator coming to *special estimator* or *discriminants* oriented towards failure as, for example, the kurtosis for rolling element bearings. The, passing to the *function estimators* of vibroacoustical processes, we will show a strong connection of the frequency spectrum of vibroacoustical processes generated by the machine and its kinematics. Finally, we will clarify the problem of the special function derived from spectra such as spectrum and multiple spectrum. The whole consideration of this point will be concluded with a short clarification of the frequency sensitivity of the different measures. In the considerations of this point and others there will be no need to indicate the dependence of the diagnostic signal on the place *'r'*; furthermore, we will accept the assumption the the tribovibro-acoustical signal has been subject to one of the methods of preliminary selection. In this connection we will finally simplify the notation of diagnostic process since we will accept $c\,(\mathrm{r}, t, \theta) \rightarrow u\,(t, \theta)$.

3.5.2 Amplitude of noise and vibrations in the diagnostics of simple machines

The common application of vibroacoustical processes in measurements have three measures of amplitude bearing the names peak value, root-mean-square value and average value. At the same time, it is obvious, owing to the random component in the observed vibroacoustical process, that each of the measures is averaged on an adequate interval of dynamic time t with interval T, called the observation time. Definitions of these measures written for process $u\,(t, \theta)$ are as follows:

average value:

$$\bar{u}(\theta) = \left[E_t \left|u(t,\theta)\right|\right] = \frac{1}{T}\int_0^T \left|u(t,\theta)\right| \; dt = u_{AV}(\theta) \tag{3.42}$$

root-mean-square value:

$$\tilde{u}(\theta) = \sqrt{E_t\left[u^2(t,\theta)\right]} = \sqrt{\frac{1}{T}\int_0^T \left|u^2(t,\theta)\right| \; dt} = u_{\mathrm{rms}}(\theta) \tag{3.43}$$

peak value:

$$\hat{u}(\theta) = E_n\left[\max\left|u(t,\theta)\right]\right. = \frac{1}{\mathrm{N}}\sum_{i=1}^{N}\left|u\left(t_{i\,\max}, \theta\right)\right| = u_{\mathrm{peak}}(\theta) \tag{3.44}$$

$$0 \le t \le T$$

The dynamics of vibroacoustical phenomena are frequently significant, reaching, for example, the order of seven decades (10^7) for noise. Then, it is convenient to use the so-called amplitude levels of the phenomenon measured in decibels defined as below:

$$L = 20\lg\left(\frac{u}{u_u}\right)\mathrm{dB} \tag{3.45}$$

where u can be an optional measure of amplitude defined above, while u_u is a conventional *amplitude of reference.*

Reference amplitudes recommended by the International Standards Organization (ISO) are the following.

(1) Reference amplitude of acoustic pressure in measurement of noise: $p_u = 2.10^{-5}$ Pa (Nm^{-2})
(2) Reference amplitude of vibration displacement: $x_u = 1.10^{-11}$ m
(3) Reference amplitude of vibration velocity: $v_u = 5.10^{-8}\ \mathrm{ms}^{-1}$
(4) Reference amplitude of vibration acceleration: $a_u = 1.10^{-5}\ \mathrm{ms}^{-2}$

The expression of vibration amplitudes with the help of decibel levels has been transferred to machine diagnostics from noise measurements, and in the manifestation of the *trend* of vibration amplitudes it is rarely met (here, trend is understood to be a tendency manifested with the increasing lifetime θ of an object). The only known diagnostic standard expressed in decibels of vibration velocities is the Canadian Standard CDS/MS/NVSH-1-A0-2, used in the Canadian Navy for the vibrational evaluation of the technical condition of deck machines.

In this way we are already in the field of machine diagnostics and, according to the title of this subsection, in the diagnostics of *simple machines*. By a simple machine we understand here the absence of functional and constructional complexity characteristic of machines designed for pressing, suction, drive, etc. A compressor, pump, fan, electric motor, for example, fulfil the conditions of a simple machine according to the above definition while, for example, a turbine set, toothed gear, turbocompressor (depending on the purpose of diagnosis) can or cannot fulfil them. On the other hand, we certainly cannot include as simple machines a gas turbine and, particularly, an aircraft jet turbine engine where, next to the constructional and functional complexities, we have also a great density of subassemblies and high flexibility of the supporting structure.

In our endeavour to formulate vibroacoustical diagnostic values describing the condition of machine, we pay attention to the fact that, in the majority of machine examination cases, the diagnostic procedure is carried out on two levels. The first question which we ask in reference to the diagnostic measurement values is: What is the general technical (utilizational) condition of the machine? This can be expresses in other words: What is the summary advancement of the wear process? For simple machines this question is often the only one. If, however, the space of machines failures is multidimensional, then the second level of diagnostics consists of a set of questions referring to the advancement of each failure separately. The answers are obtained with the help of failure-oriented measurements about which we are going to talk in detail in Sections 3.5.2 and 3.5.4.

Considering further machines of simple construction and function, we can recall the example of the tribovibroacoustical model of machine (Section 2.3.6) where the mutual transformations of the energy of wear processes into vibroacoustical ones were taken into consideration. As we know, the measure of a *stream of* kinetic *energy* of the vibroacoustical process is the product of the density of medium and the square of vibration velocity ($1/2\ \rho\ v^2$).

Therefore, the vibration velocity, according to this model, should reflect in the best way the general advancement of the wear processes and thereby the technical consideration of a simple machine. Let us see, then, how these indicators agree with the diagnostic practice. The evaluation of the utilization condition of a pump unit with the help of noise measured in the vicinity of the machine and the velocity of vibrations measured on the bearing of the electric motor and pump are presented in Fig. 3.21 [67]. As we see, noise can also carry information about the condition of a machine, but its sensitivity is very small (maximum 5dB) which agrees with our conclusions about the detection of failures of a mechanical nature (Fig. 3.38). However, it is worth remembering this fact in the estimation of the advantages of diagnostics in reference to environmental protection (Table 1.2).

The peak amplitude $\hat{v}$ of vibration velocity measured on the bearings of the motor and pump is here, as we see, a significantly better indicator of the pump condition, since

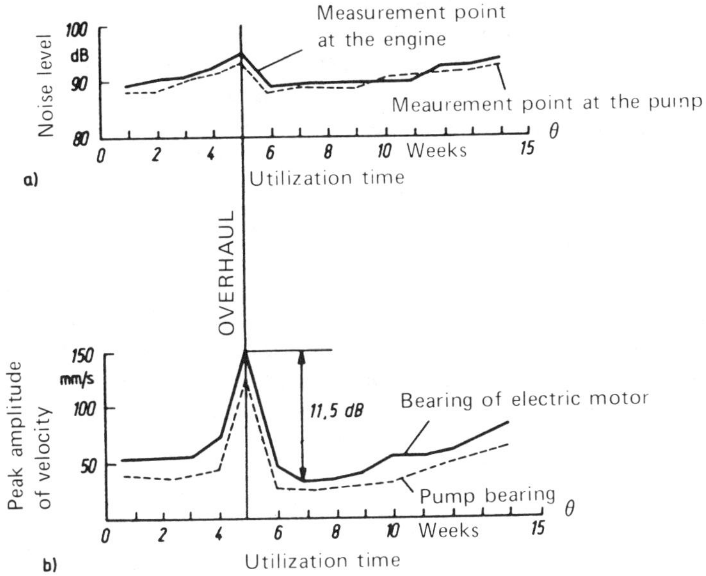

Fig. 3.21 — Trend of change in noise levels (a) and in the vibration amplitudes (b) in a pump [67]

the difference of amplitudes before and after the overhaul is more than threefold (11 dB); furthermore, this quantity is significantly less susceptible to external disturbances than noise, giving a greater guarantee of a *reliable diagnosis*. Let us notice, also, the exceptionally great amplitudes of vibrations (150 mm/s) being proof of the elastic mountings of the unit and the great participation of vibration of the entire machine as a rigid solid body. How far the absolute value of vibration amplitude can be misleading is confirmed by Fig. 3.22, taken from materials of the same American diagnostic company, IRD [68}. Here, the maximum vibration amplitude reaches 15 mm/s, and a greater dynamic change on the motor bearing indicates the cause of failure.

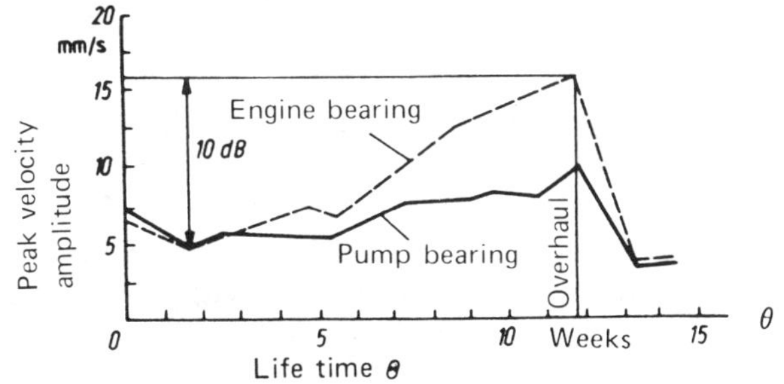

Fig. 3.22 — Course of utilization changes in the level of vibration velocity of bearings in a pump [68]

From the above two examples, it follows that the Anglo-Saxon countries prefer the peak velocity $\hat{v}$ as the diagnostic value which agrees also with the theoretical considerations the Author has quoted [39].

A further conclusion following on from this data is that in diagnostics the relative change of vibration amplitude from the rated level is important, and that it must be at least twofold $[\hat{v}(\theta)/\hat{v}(0) \geq 2$ (i.e. by $20 \lg 2 = 6$ dB) in order to tell us about a change in the technical condition. In each case shown in Figs 3.21 and 3.22 the overhaul took place when $[\hat{v}(\theta)/\hat{v}(0) \cong 3 \rightarrow 10$ dB. Such regularity is visible also in the Anglo-Saxon standards of diagnostics (included in the Appendices), where the qualification of the change in the condition takes place every 6–10 dB, while the range of possible amplitudes for a new machine and a machine with failure reaches from 20 to 30 dB. As an example of such standards, Fig. 3.23 shows the guidelines of the PMC/BETA Corporation [69] determining the warning ranges of vibration amplitudes of different machines. The entrance into this range (marked by a rectangle) determines the beginning of the period of accelerated wear where it is recommended to detect the reason of failure, e.g. by the spectral analysis of vibration composition and the preparation of the necessary range of overhaul.

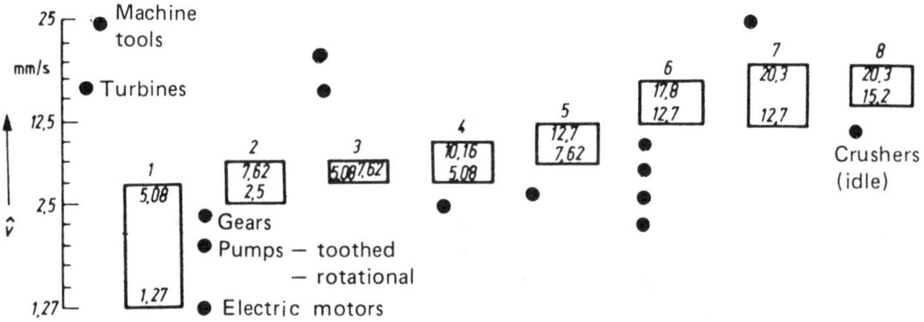

Fig. 3.23 — Guidelines of the warning ranges of vibration amplitudes of different machines according to PMC/BETA [69]

In European countries, on the other hand, it has been accepted, according to the guidelines of ISO, to evaluate the utilization condition of machines by the measurement of the root-mean-square amplitude of velocity $\tilde{v}$ where the change of the condition class is determined by the vibration difference of 8 dB, and the range between the classes Good and Inadmissable is 16 dB.

As we see from Fig. 3.24, the dynamics of changes in the root-mean-square value of velocity is smaller; on the other hand, the absolute amplitude of velocity reaches from 0.71 mm/s for the Good class of small machines to 18 mm/s for the Inadmissable class of heavy machines.

To conclude the discussion on the diagnostic guidelines for simple machines, it is worth comparing the maximum values of the different standards in the Inadmissable class. They are the following (see also the Appendices):

British diagnostics company VCI Ltd	$\tilde{v} = 15.24$ mm/s	
American diagnostic company IRD	$\hat{v} = 6$ mm/s	
ISO guidelines 3945 or 2372	$\tilde{v} = 18$ mm/s rms	
PMC/BETA guidelines	$\hat{v} = 20$ mm/s	(3.46)
Canadian government CDMS/MS guidelines	$\tilde{v} = 31.4$ mm/s rms	
Blake's guidelines/service factor D = 1, generally 0.3–3	$\hat{v} = 50$ mm/s	

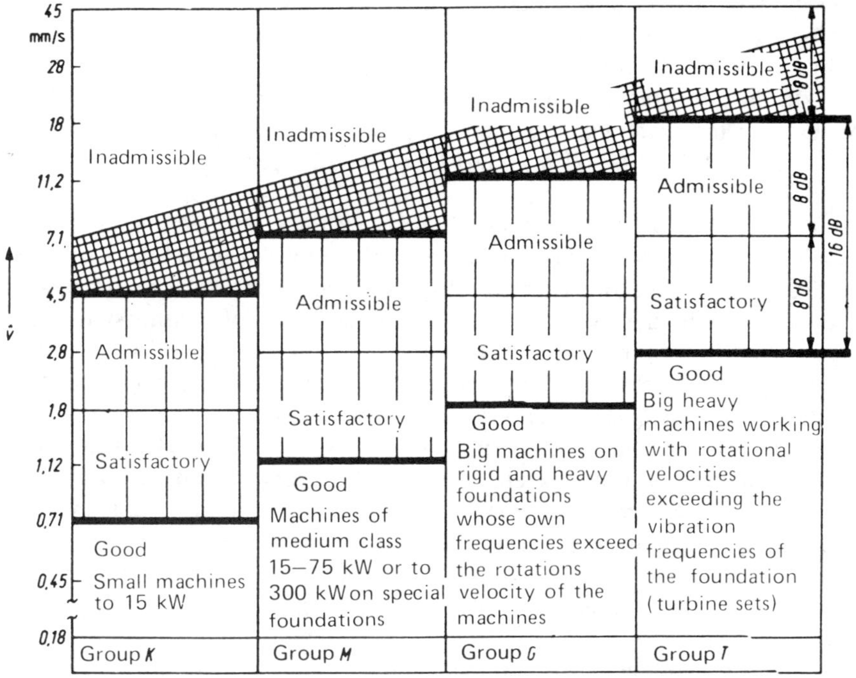

Fig 3.24 — Classification of vibration condition of four groups of turbomachines for working rotations from 10–120 rev./s (according to ISO-3945 or ISO-2372)

As we can see from the above, next to the confusion of the criterion values (root-mean-square amplitude $\tilde{v}$, peak amplitude $\hat{v}$) we have a several-fold span of values within the same criterion value. On the other hand, the quoted cases of pump diagnostics (Fig. 3.21) is covered only by Blake's guidelines and with an availability factor of 3. Hence, we see also that the determinations of the standards of diagnosis is risky, and it is better to limit the diagnosis to one definite industrial plant, which will be discussed in Chapter 4.

Is it always worth measuring the vibration velocity and not the displacement or acceleration? The majority of industrial machines—compressors, turbines, pumps and even industrial transmission gears—work with journal bearings. In such cases, the peak amplitudes of relative displacements of shaft and bearing bushing are limited by measuring them with the help of contactless transducers. The admissible peak amplitudes of vibrations according to the American Petroleum Institute for steam turbines are given in API-611; others are as follows.

(1) For special steam turbines, API–612.
(2) For industrial transmission gears, API-613.
(3) For gas turbines, API-616.
(4) For rotodynamic centrifugal compressors, API-617 they are:

$$x_l = 25.4 \sqrt{\frac{12\,000}{n}} \ \mu\text{m} \ (\text{maximum } 50.4 \ \mu\text{m}) \tag{3.47}$$

where n is in revolutions per minute (these standards were originally given in inches, 25.4 μm).

In cases of other machines with journal bearings the recommendation of the manufacturer should be consulted, or where no such data are available, the recommendations of the German Engineer's Association VDI-2059 in the corresponding parts 2 + 5 should be applied. In all these recommendations the admissible amplitude of relative vibrations decreases with the revolutions ranging for slow-speed machines from $\hat{x}_l = 320\ \mu m$, according to VDI-2059 (1000 rev./min) to $\hat{x}_l = 12.7\ \mu m$ according to API at 48.10^3 rev./min.

At the conclusion of the discussed turbomachines and similar ones it is worth mentioning that the vibrations of the bodies of these machines and the bearing housings are evaluated according to the already discussed limitation of the root-mean-square, or peak velocity of vibrations with the exception of the bodies of transmission gears and the rolling element bearing housings. Here, due to the discrete way of load transfer (teeth, rolling elements) the proper working frequency of load transfer is many times higher than the rotation frequency f_0 (see Sections 2.2.3–2.2.6), exceeding in many cases the limit of an effective measurement of the vibration velocity 1000–2000 Hz. Thus, the general technical condition of bearings and transmission gears should be evaluated by the measurement of acceleration of the housing or body. Among the recommendations concerning this problem it is worth mentioning, for example, the guidelines of the American Gear Manufacturers' Association (AGMA) which state that, for $f > 500$ Hz, the acceleration amplitudes of the gear body cannot exceed $10g$ ($g = 9.81\ m/s^2$) for durability reasons.

Guidelines for rolling element bearings are given by Blake in his monograph [70] in categories of three measurement values: (a) the peak value of acceleration amplitudes; (b) the oscilloscope observation of this value; and (c) the peak value of the vibration velocity of the housing, as in Table 3.1. Such three parameter observation of the bearing permits 95% accuracy of diagnosis.

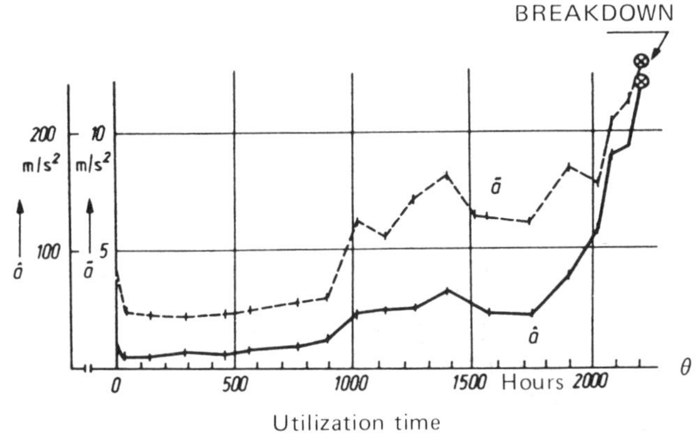

Fig. 3.25 — Changes in the root-mean-square amplitude $\tilde{a}$ and the peak acceleration amplitude $\hat{a}$ of an electric motor measured in the function of the utilization time [71]

Concluding the discussion of the techniques of vibrational diagnosing of simple machines, let us return to the topic of peak value/root-mean-square value. Arguments for the peak value in the diagnostics in electric motors are presented in Fig. 3.25 quoted from a master's thesis [71]. As we see from the figure, the dynamics of change in the peak

Table 3.1 Qualification of vibration symptoms of rolling element bearing condition [69] (30–200 mm diameter)

Failure classification–condition		Peak value of housing vibration acceleration		Peak value of housing vibration velocity
		$\hat{a}-g$	Impulses g	$\hat{v}$ mm/s^{-1}
AA	Catastrophic (breakdown, shutdown)	100	200	140.5
A	Dangerous (repair within 2 days)	17.8	35	59.25
B	To planned replacement (within 21 days)	3.16	6	25.00
C	Small defect	0.56	1	10.5
D	Without failures	0.1–0.01	0.2	4.5–0.25

$g = 981$ cm/s^{-2} PEAK

Impulses → instantaneous, high, over the mean level observed by oscilloscope

values are twice as great as peak amplitudes; similarly, the readability of change in the state of this diagnostic value.

Summing up the vibroacoustical evaluation techniques of the condition of simple machines, we can say that when we went to evaluate the condition of the rotor assembly and its unbalance, we will measure the displacements of relative vibrations of the shaft in relation to the bearing bushing. In the remaining cases we will measure the vibration velocity amplitude of the machine body or the bearing with the exception of rolling bearings and transmission gears, where it is recommended to measure the peak accelerations of the bearing housing or transmission gear body.

3.5.3 Special point estimators of vibroacoustical processes

The definitions of amplitudes of vibroacoustical processes given in the previous chapter are, in fact, the moments of the modulus of the stationary and ergodynamic random process $u\ (t)$ (see, for example, Bendat and Piersol [53]). This way of looking on the measurement of the vibroacoustical process amplitude can be generalized by obtaining an amplitude of the q-th order as the q-th root from the moment of the q-th order of the vibroacoustical process [71]:*

$$\bar{u}_q = \left[\int_{-}|u|^q \, p(u)\, du \right]^{1/q} = \left[\frac{1}{T} \int_0^T |u(t)|^q \, dt \right]^{1/q} = \sqrt[q]{E_t |u(t)|^q} \qquad (3.48)$$

where $p\ (u)$ is the density of amplitude distribution probability of the $u\ (t)$ process, while

* In this section, for the sake of convenience, we are going to omit time θ in the determinations of the $u\ (t)$ process.

the definition in the domain of time is important for the ergodic and stationary processes.

Looking at the above definition in the time and amplitude domains and the definitions in Equations (3.42)–(3.44) we can find that for $q = 1$ we have an average amplitude in Equation (3.42), for $q = 2$ a root-mean-square amplitude in Equation (3.43), and for $q \to \infty$ the peak amplitude in Equation (3.44) [7, Section 3.9.1].

It can be concluded from the data of diagnostic supervision and from the tribovibroacoustical model presented in the previous chapter, that the problem of the absolute value of the input level of vibrations in a new machine or in a machine after overhaul is always controversial. Leaving this problem aside, the numerical dimensionless measures of vibroacoustical processes have been defined and they can be generally expressed in the following form:

$$\zeta = \frac{\left[\int_{-} |u|^{q} \; p(u)\, du\right]^{1/q}}{\left[\int_{-} |u|^{m} \; p(u)\, du\right]^{1/m}} \tag{3.49}$$

The significance of the general dimensionless estimator for some q and m values is the following:

$$\left.\begin{aligned}
K &= \tilde{u}/\bar{u} = \text{form factor for } q = 2, \quad m = 1 \\
C &= \hat{u}/\tilde{u} = \text{peak factor for } q \to \quad , \quad m = 2 \\
I &= \hat{u}/\bar{u} = \text{impulse factor for } q \to \quad , \quad m = 1 \\
L &= \hat{u}/u_{1/2} = \text{clearancefactor for } q \to \quad , \quad m = 1/2 \\
\sqrt[4]{\beta} &= \text{the fourth root of kurtosis } \beta \text{ for } q = 4, \; m = 2
\end{aligned}\right\} \tag{3.50}$$

The form factor K has a very small sensitivity to condition changes; on the other hand, the clearance factor L is difficult to measure for $m = 1/2$. The remaining factors find an application in the diagnostics of rolling bearings and transmission gears. Let us first consider the peak factor C, which is easy to obtain by measurement. A comparison is presented in Fig. 3.26 of the diagnostic sensitivity of the acceleration peak factor C_a velocity C_v and vibration displacement C_x observed on the rolling bearing housing of an electric motor [71].

As we see in Fig 3.26, the peak factor of displacement C_x has no diagnostic value, C_v has a small diagnostic value, while C_a is a very good indicator of the technical condition of the bearing, and it could be proposed to assume $C_a = 10$ as limit of the GOOD CONDITION of the bearing. For rolling element bearings with greater overall dimensions and greater load, the peak factor of velocity C_v has also a diagnostic importance. Recently, Andrzejewski [72, 73] reported two cases of a correct assessment of the motor bearing in a pump when he used the critical value $C_v = 2$ (Fig. 3.27).

The impulse factor I has similar diagnostic properties and, furthermore, is characterized by a greater sensitivity. A confirmation of this fact can be found in Fig. 3.28 from the investigations on the electric low-power motor. Here, for a better discrimination of changes and the elimination of disturbances, a preliminary filtration was applied with

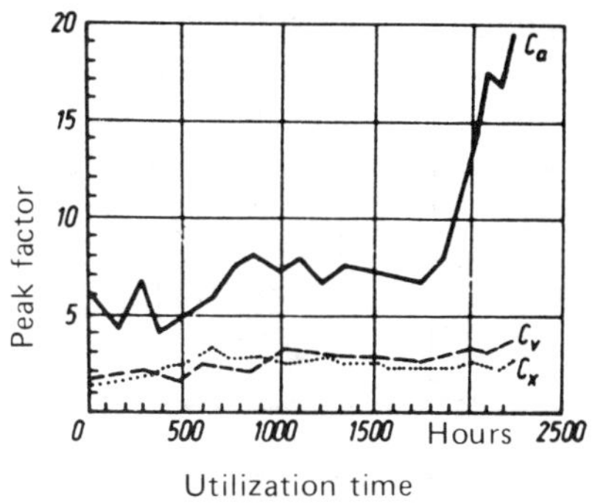

Fig. 3.26 — Utilization trend of changes in the peak factors of acceleration C_a, velocity C_v and vibration displacement C_x of the bearing housing of an electric low-power motor [71

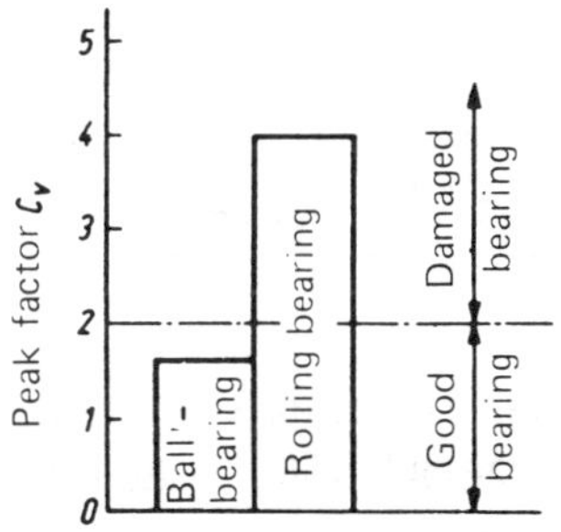

Fig. 3.27 — Peak factor of vibration velocity C_v for the ball bearing and roller bearing of an electric high-power motor [72]

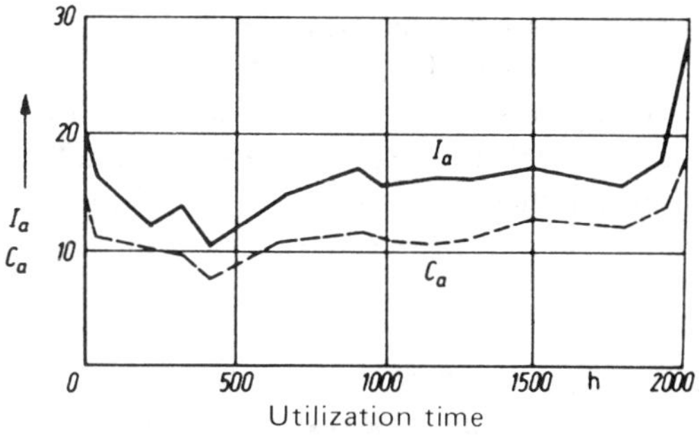

Fig. 3.28 — Utilization trend of the impulse factor I_a and the peak factor C_a in the octave 8 kHz for the vibrations of the bearing housing of an electric low-power motor

the help of an octave filter of 8 kHz.

The preliminary filtration is also used with success in the measurement of kurtosis β in the evaluation of the condition of rolling element bearings. As indicated by Dyer's [74] investigation, the value of kurtosis β of new bearings is contained within the limits of β = 3–4, while for impaired bearings β = 10–20. This regularity is maintained within wide limits of rotations and overall bearing dimensions. Comparative behaviour of the

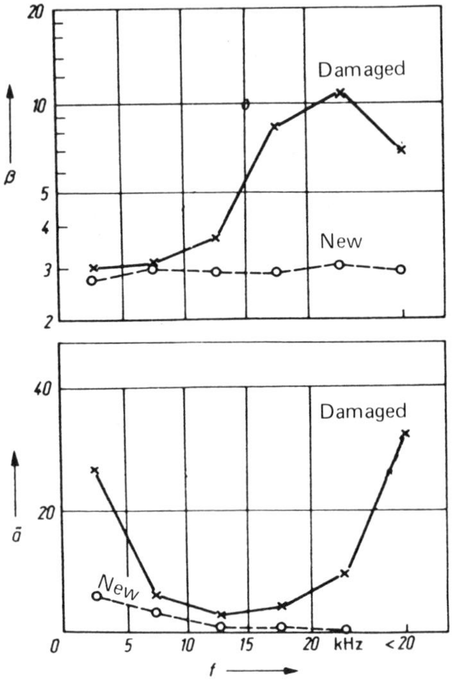

Fig. 3.29 — Comparison of the value of kurtosis β and the root-mean-square acceleration $\tilde{a}$ of the vibrations of a bearing housing in six bands with $\Delta f = 5$kHz [75]

amplitude of acceleration and its kurtosis for the vibrations of bearing in five bands $\Delta f =$ 5 kHz and the sixth (final) band $\Delta f < 20$ kHz are shown in Fig. 3.29.

As we see from the figure, the value of kurtosis β in the sixth band for $\Delta f < 20$ kHz and the damaged bearing decreases significantly, while the bands $\Delta f = 5$kHz with 17.5 kHz and 22.5 kHz still have high values of kurtosis. It indicates a significant role of of filtration in diagnostic investigations, since the definition of kurtosis is based on the density of the distribution of $p\,(u)$. Therefore, we can say that in the band $\Delta f > 5$ kHz the density of $p\,(u)$ distribution and, therefore, also of kurtosis β loses its properties of being oriented to the failure and approaches the small values characteristic of Gaussian-like distributions.

This is a very abbreviated account of the special estimators defined in the field of the amplitude of the vibroacoustical process. More information can be found in the Author's monograph [7] and in more recent scientific and applied works.

Point estimators can also be defined in the field of frequency, as shown in the Author's work [76] the problem is developed in his monograph [7]. Let us quote two spectral estimators: (a) Rice's frequency of $u\,(t)$ process determined by f_u ; and (b) the indicator of harmonicity H_u. If, next to the original process $u\,(t)$, we define the differential process $u^{(1)}(t) = \dfrac{du}{dt}$ and the integral process $u^{(-1)}(t) = \int_0^t u(\tau)\, d\tau$, then the above-mentioned estimators can be defined in the following way:

$$f_u = \frac{1}{2\pi}\left[\frac{\int\limits_0 f^2 G_{uu}(f)\,df}{\int\limits_0 G_{uu}(f)\,df}\right]^{1/2} = \frac{\left(\tilde{u}^{(1)}\right)}{2\pi\,\tilde{u}}$$

$$H_u = \frac{f_u}{f_u^{(-1)}} = \frac{\left(\tilde{u}^{(1)}\right)\left(\tilde{u}^{(-1)}\right)}{\tilde{u}^2} \tag{3.51}$$

where ($\tilde{\cdot}$) denotes the root-mean-square value of the process in brackets.

As we can see from the notation of Equation (3.51), these values can also be calculated by measuring to root-mean-square amplitude of the integrodifferential processes, e.g. u (t). Definitively speaking, if the vibration velocity $u\ (t) = v\ (t)$ is the original process, then by measuring the acceleration $a\ (t) = u^{(1)}(t)$ and the displacement $x\ (t) = u^{-1}\ (t)$ we can calculate the values of Rice's frequency and the indicator of harmonicity. The physical interpretation of Rice's frequency indicates an average frequency of the band, and if $\Delta f \rightarrow 0$ as for the harmonic process, then f_u is its real frequency. The harmonicity indicator, on the other hand, indicates dimensionless the width of the band occupied by the process, since for a harmonic $H_u = 1$ and for white noise $H_u \rightarrow \infty$.

The diagnostic application of these values can be different since, without a spectral analysis, we can obtain information about the average frequency of the vibration process f_v and the width of the occupied band H_v when we measure the root-mean-square values of the displacement $\tilde{x}$, velocity $\tilde{v}$, and acceleration $\tilde{a}$. Then: $f_v = [\tilde{a}/2\pi\tilde{v}\ )]$, $H_v = [\tilde{x}\ \tilde{a})/\tilde{v}^{\ 2}$. For example, only from the measurement of $\tilde{v}$ and $\tilde{a}$ can one judge without spectral analysis whether the recorded increase of vibration amplitudes occurred as a result of the increase in unbalance, if the resulting $f_v \sim f_0$, or whether, for example, it occurred because of damage to the bearing, transformation gear, etc. if $f_v \gg f_0$ (where f_0 us the rotary frequency of the machine). These estimators can also find an application in the diagnostics of rolling bearings, which is proven by Fig. 3.30 where, with the deteriorating condition of the rolling element bearing, both values behave in a similar way.

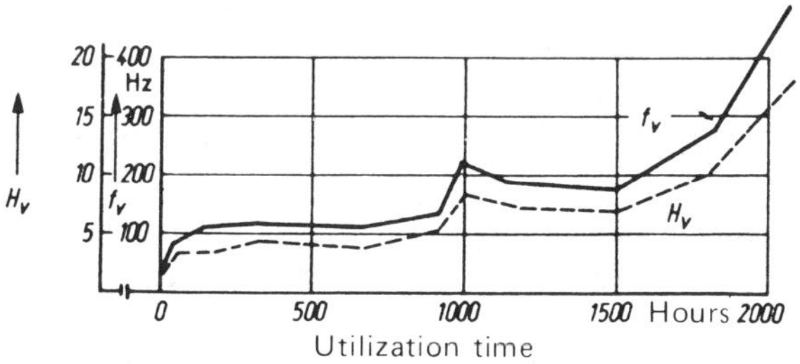

Fig. 3.30 — Frequency discriminants f_v , H_v for the vibrations of the bearing housing of an electric low-power motor [71]

We see here that the average frequency of the process grows from the frequency of the magnetostriction of motor housing ($f_v \cong 100$ Hz) showing, first, microfailures in the bearings for $\theta = 1000$ h ($f_v \cong 200$ Hz) their number increasing with the increase of θ. The indicator of harmonicity H_v grows here similarly from $H_v = 5$ to $H_v = 20$, indicating a fourfold widening of the band of the motor's vibration process owing to the failure of the roller bearing.

Above, we have defined and shown the application of special measures of the vibroacoustical processes, and at the same time some of them, such as kurtosis β, have a failure-oriented character. Dimensionless estimators, such as the peak factor c, impulse factor I, kurtosis β and the index of harmonicity H_u, have special importance, since their rated levels refer to the condition of the object and not to its type, e.g. β = 3–10 for bearings in good condition independent of their place of installation.

However, the complexity of vibroacoustical signals is not always so small as in the above cases where the band filtration of Rice's frequency permits us to separate the tribovibroacoustical signals of the investigated subassembly. Frequently, for the same purposes we have to apply far more subtle measures than Rice's frequency of the process, and they have to be connected with the functioning mode of the object. We will try to clarify this problem in the following section referring to the spectral (frequency) analysis of vibroacoustical processes.

3.5.4 Spectral and kinematic analyses in vibroacoustical diagnostics

The second level of diagnostic investigations is undertaken when the machine enters the period of accelerated wear, which refers to the localization and determination of the degree of advancement of the particular failure. Such possibilities for simple machines are created by linking the analysis of vibration frequency with the kinematics of the machine or assembly (Chapter 2). However, before we pass to this subject, let us note the fact of a natural connection between the root-mean-square amplitude of the process $\tilde{u}(\theta)$ and its spectral density of power $G_{uu}(f, \theta)$. From Parseval's theorem [77], there follows the relation:

$$\tilde{u}(\theta)^2 = \frac{1}{T}\int_0^T u(t,\theta)^2 \, dt = \int_0 G_{uu}(f,\theta)\, df$$

$$\text{where } G_{uu}(f,\theta) = E_N\left\{\frac{\left|U_i(f,\theta)\right|^2}{T}\right\}, \tag{3.52}$$

$$U_i(f,\theta) = \int_0^T k_i(t,\theta)\exp(-2\pi j f t)\, dt,$$

$u_i(t, \theta)$ are the T-segmental realizations of process $u(t, \theta)$ with the amount $i = 1,\ldots, N$; $E_N(\cdot)$ is the averaging operator of this set, $U_i(f, \theta)$ is Fourier's realization transform, and the dimension of spectral density is $[G_{uu}(f, \theta)] = [u]^2/\text{Hz}$; $j = \sqrt{-1}$.

It follows from the above that the root-mean-square amplitude is a sum (integral) of all spectral contributions from each elementary component with frequency $\Lambda f \varepsilon (0, \infty)$.

In Chapter 2, we discussed vibroacoustical phenomena, a summary of this state of knowledge is presented in Table 3.2, supplying the most important components of tribovibroacoustical processes in relation to rotational frequency f_0. Further and more detailed data can be found in the diagnostic chart included in the appendix.

The method of using the knowledge contained in the table and the idea of the summation of vibrational contributions according to Equation (3.52) are shown in

Table 3.2 Spectral components of vibrations associated with the occurrence of failures in machine elements in comparison with rotation frequency [7]

	f_0, rev./s or Hz		
Defect description	Dominating vibration frequency	Vibration direction	Remarks
Unbalance of rotating element	$1 \times f_0$	Radial	Common cause of vibrations in machines Amplitude proportional to unbalance
Misalignment or bending of shaft	$1 \times f_0$ usually $2 \times f_0$ $3 \times f_0$	Radial-axial	If amplitude of component $2f$ (axial) exceeds 75% of f_0 component, then such failure may cause breakdown of machine
Damaged rolling bearing element	Impulses with frequencies connected with rotations and type of impaired element Also vibrations in high frequency band (20–60kHz)	Radial and axial	Impulse frequency connected with defect: β = angle of operation n = number of rolling elements f_r = relative rotation frequency of raceway in hertz (internal raceway in relation to external)
	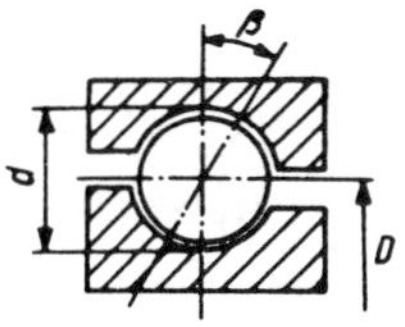		Rotation frequency of raceway or shaft $f = f_r$ d = rolling element diameter D = pitch diameter Defect of external raceway $f = \frac{n}{2} f_r \left(1 - \frac{d}{D} \cos\beta\right)$ Hz Defect of internal raceway $f = \frac{n}{2} f_r \left(1 + \frac{d}{D} \cos\beta\right)$ Hz Defect of one rolling element $f = \frac{D}{d} f_r \left[1 - \left(\frac{d}{D}\right)^2 \cos^2\beta\right]$ Hz Runout, clearance of cage $f = \frac{1}{2} f_r \left(1 - \frac{d}{D} \cos\beta\right)$ Hz

Table 3.2 (continued) Spectral components of vibrations associated with the occurrence of failures in machine elements in comparison with rotation frequency [7]

f_0, rev./s			
Defect description	Dominating vibration frequency	Vibration direction	Remarks
Clearance in slide bearing settling	$1/2\,f_0$ $1/3\,f_0$ subharmonics of rotations	Radial	Such clearances are revealed at determined rotations only and at determined temperature, e.g. in turbogenerators
Instability of oil film in slide bearing (oil vortex, resonance hysteresis vortex)	From 0.20 to $0.48 \times f_0$	Mainly radial	Always less than $1/2\,f_0$ frequently occurs in high-speed machines and turbines
Damaged or toothed gears	$n \times f_0$ $z \times f_0$ and harmonics: b number of revolutions z number of teeth	Radial and axial	Sidebands around meshing frequencies indicate eccentricity of toothed wheels (detectable only with the help of narrow-band analysis $f =$ const)
Clearances in rotary pairs, clearance in foundation	$1 \times f_0$ $2 \times f_0$	Radial or axial	Usually in connection with unbalance and lack of alignment
Impaired driving belt	$1 \times f_0$ of belt $2 \times f_0$ of belt $3 \times f_0$ of belt $4 \times f_0$ of belt	Radial	This failure is easily detected by stroboscope
Unbalanced forces and moments in to-and-fro motion	$1 \times f_0$ and higher harmonics depending on the order of unbalance	Mainly radial	Inseparably connected with machine operation They can be decreased only by construction changes or vibroisolation
Pulsation of working medium	$f_0 \times$ number of blades, cylinders, etc. and higher harmonics	Radial and axial	In case of coincidence with resonances of construction they can cause failure
Vibrations of alternating current electric machines	1 or 2 $\times$ power supply frequency in electric network	Radial or axial	They should disappear after switching off the supply current

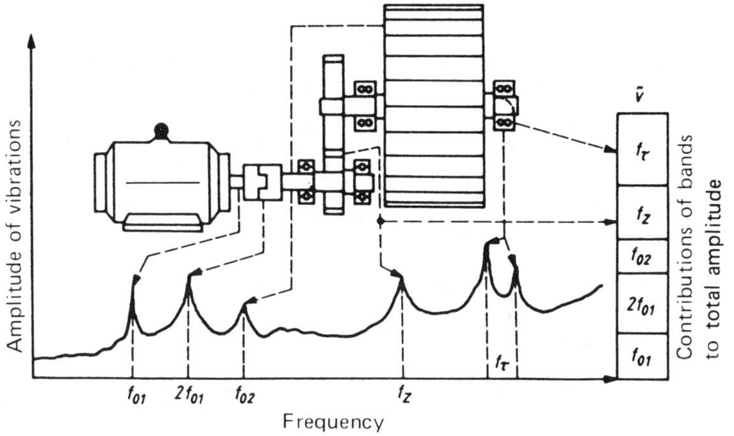

Fig. 3.31 — Spectral image of the vibrations of simple machine elements and spectral contributions to the total level within the linear range, Lin

Fig. 3.31 where almost each rotating element of a simple machine has its spectral image and contribution to the total amplitude of root-mean-amplitude velocity $\tilde{v}$ (right-hand side if the figure).

In this way, a fusion of the knowledge from the field of kinematics and machine dynamics takes place, subordinating to each rotating elements its due spectral line which allows identification of the advancement of the wear process. Such identification is almost obvious from the data contained in Fig. 3.32 [68] where the almost advanced failure after 11 weeks of operation consists of the slackening and misalignment due to a strong component $2f_0$.

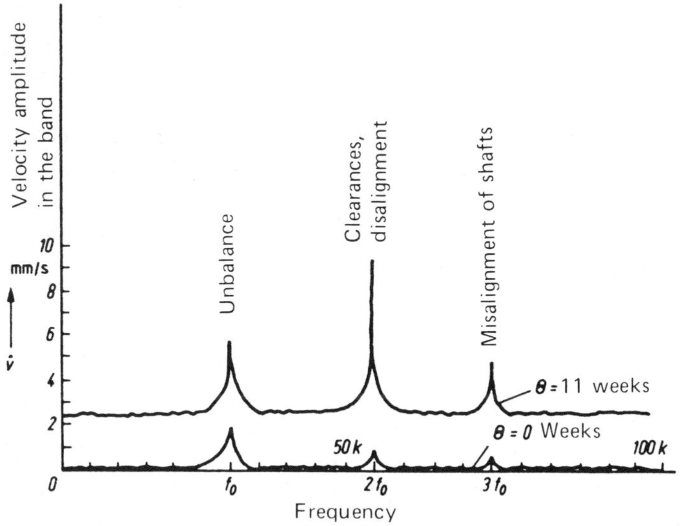

Fig. 3.32 — Change of the spectrum of fan vibration in use with visible evolution: unbalance f_0, slackness in joints $2f_0$, no alignment of motor and fan shafts $2f_0$, $3f_0$ [68]

The two drawings in Fig. 3.1 in connection with the conception of the life curve of a

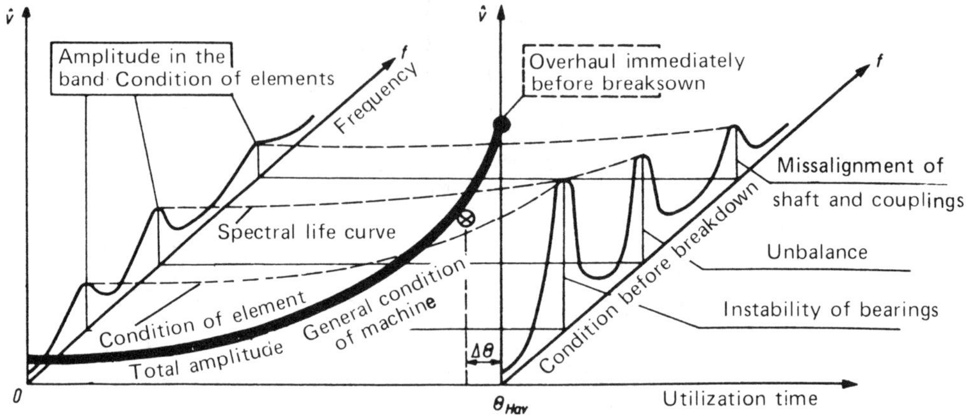

Fig. 3.33 — Illustration of the idea of vibrational diagnostics of simple machine by measurement of the amplitude of total vibration velocity for the evaluation of the general condition, and by spectral analysis of vibrations for the evaluation of the advancement of the particular failures (global and spectral life curves)

machine in the fields of frequency f and life time of the object θ, as in Fig. 3.33. As we see, the figure, contains the whole idea of the vibrational diagnostics of simple machines where, for the evaluation of the general technical condition, it is necessary to measure the peak value of vibration velocity. On the other hand, for the localization of failure and the evaluation of its advancement, we must observe the partial curves of the corresponding spectral components. So we see that by the measurement we are able to *diagnose, find the origin and determine the prognosis of the condition of the machine and its subassemblies.*

The spectral analysis of the vibrations of simple machines in connection with the kinematic analysis can render valuable service. Hence, it is worth having a closer look at the instrumental features of spectral analysis as an experimental approximation of Equation (3.5). Four types of analysis of the same phenomenon in the band 0 : 1500 Hz are presented in Fig. 3.34 [78]: (a) with the help of an octave analyser $[(\Delta f)/f] = 0.71$; (b) terce analyser $[(\Delta f)/f] = 0.231$; (c) narrow-band analyser with a constant relative band width $[(\Delta f)/f] = 0.08 = 8\%$; and (d) narrow band analyser with a constant band width $\Delta f = 2$ Hz (analyses of the same type can be obtained also by computer). As we see from Fig. 3.34, the octave analysis is almost useless, similarly analysis for higher frequencies. In turn, the narrow-band analysis $\Delta f = 2$ Hz is too detailed, showing analysis elements in reference to which we have no possibility of a cause-and-effect subordination. The optimal analysis is here (in the majority of cases in diagnostics) the narrow band analysis $[(\Delta f)/f] = 7–10\%$ in which we see the kinematic details of the spectrum, but without the details of the subtle spectrum structure.

Concluding the considerations referring to the usability of the spectral analysis in diagnostics, it is worth analysing Fig. 3.35, which reveals failures in the frequency domain (Fig. 3.35). Generally, we can say that: (a) the band of up to several f_0 (the recommended measurement of displacement) reveals failures connected with the basic rotations; (b) the band from several f_0 to several hundred f_0 comprises failures connected with the discretization of the basic motion (of the bearing, meshing, blade pulsation); and (c) the band over several hundred f_0 is a band of tribovibroacoustical microphenomena

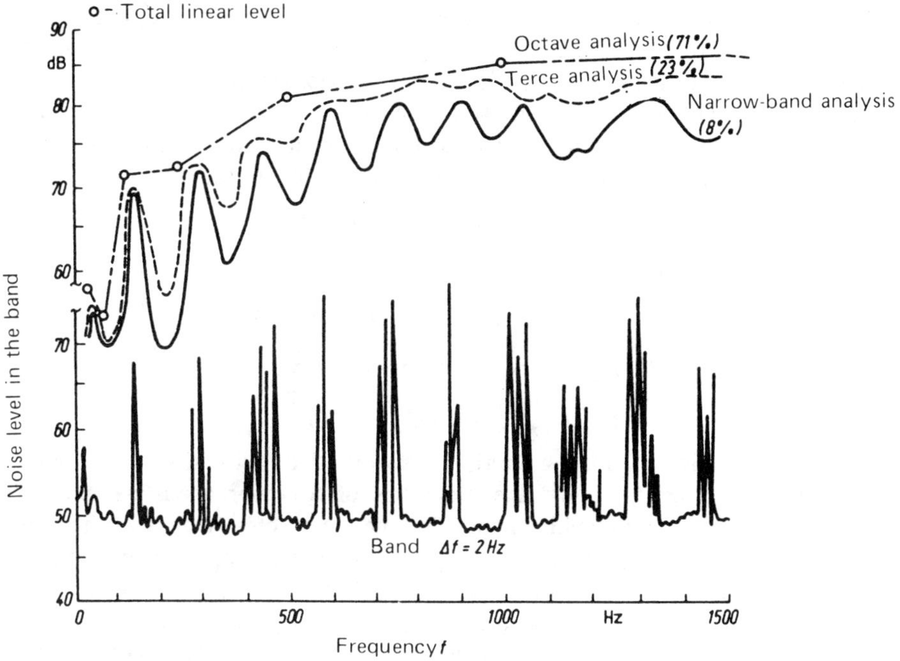

Fig. 3.34 — Spectrum of noise in rolling element bearing obtained with different analysers [78]

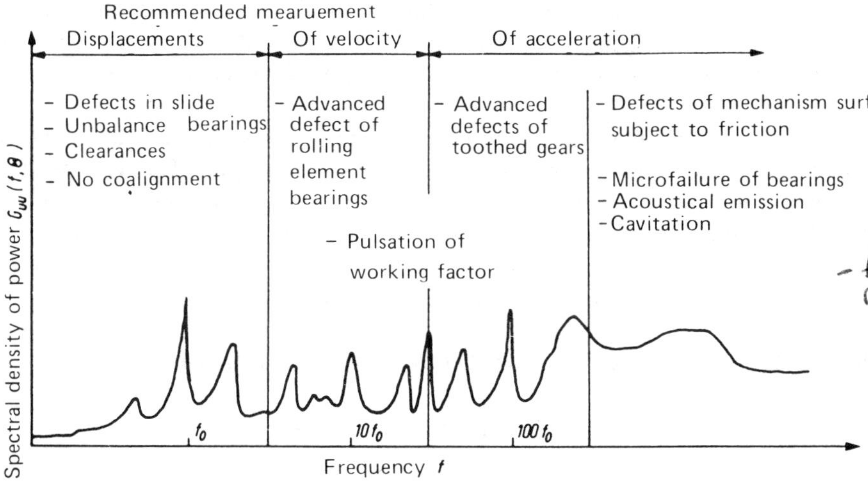

Fig. 3.35 — Spectrum of the acceleration of vibrations of a hypothetical machine with the indication of bands characteristic of different groups of failures as a function of rotational frequency f_0 [7]

(cavitation, friction, acoustical emission, etc.). It is worth keeping Fig. 3.35 in mind when we consider the selection of the observation band and the best measurement values of vibroacoustical process.

3.5.5 Special functional spectral measures of vibroacoustical processes

The main part of the vibroacoustical signal is the sum of convolutional members of propagation properties $h\ (r_i, t, \theta)$ and the forcing tribovibroacoustical process, or $V_i\ (t, \theta)$. Assuming that an ideal preliminary selection has removed the rated signal $c\ (r, t, \theta)$, noise $n\ (r, t, \theta)$ and all other tribovibroacoustical components, with the exception of the one we are looking for, then we can write the following formula for the signal"

$$u\ (r, t, \theta) = h\ (r, t, \theta) * V\ (t, \theta)$$

Its Fourier transform is:

$$(r, f, \theta) = (r, f, \theta) * V\ (f, \theta)$$

and its spectral density of power will be [7, Section 3.5]:

$$G_{uu}(r, f, \theta) = |H\ (r, f, \theta)|^2\ G_{vv}(f, \theta) \tag{3.53}$$

As we see, this quantity comprises properties of vibration propagation through transmittance $H\ (r, f, \theta)$ and the properties of tribovibroacoustical processes of the machine connected with its kinematics through the spectral density of the original signal of failure $G_{vv}\ (f, \theta)$. In many cases, the propagation path properties are almost invariable, so that the operation separating Equation (3.53) with the stressing of the kinematic properties (periodicity) of the original process of failure would be diagnostically valuable. This operation is called *ceptstrum* and its origin has been explained by the Author [7, Section 3.5] among others; here, we will give only its most important properties and diagnostic implications. In its definition it comprises two operations on the spectrum of power $G_{uu}\ (f, \theta)$ (here and henceforth we will omit the coordinate r), i.e. finding of the logarithm and Fourier's transformation. Finding the logarithm changes the product of function in Equation (3.5) into a sum; thus, it divides in this sense the properties of the propagation, and forcing it increases the dynamics of the components taken into consideration.

Fourier's transformation has the purpose to indicate the periodic components, and it can be in the simple or the inverse form, depending on the type of power spectra taken into consideration in the calculations. With unilateral power spectrum, without negative frequencies, as in Fourier's theory, it is convenient to use the simple transformation:

$$\Im[\cdot] = \int_0^{\Delta F} (\cdot) \exp(-2\pi f j\ \tau)\ df$$

while with the bilateral spectrum it is better to use the inverse transform:

$$\Im^{-1}[\cdot] = \int_{-\Delta F}^{\Delta F} (\cdot) \exp(2\pi f j\ \tau)\ df$$

where $j = \sqrt{-1}$, ΔF is the range of frequencies taken into account in the spectrum, and τ is time as an analogue of the period. The general definition of cepstrum is the following:

power cepstrum
$$C^2(\tau) = |\Im[\log G_{uu}(f, \theta)]|^2$$

amplitude cepstrum
$$C(\tau) = \sqrt{|\Im[\log G_{uu}(f, \theta)]|^2} \tag{3.54}$$

The consideration of the root in the definition of amplitude cepstrum has the purpose of preserving the components with smaller amplitudes which unfortunately are lost by raising to the second power. As we see from the definition, it is difficult to speak about physical meaning and the scale of the amplitude of cepstrum $C(\tau)$. On the other hand, time τ as argument is not a dynamic time t, nor a time of delay as in the correlation function. But it is an analogue of the basic period of a process. So, when in the course of cepstrum analysis the component of time τ_0 is visible, it denotes the periodical component with the frequency $f = 1/\tau_0$.

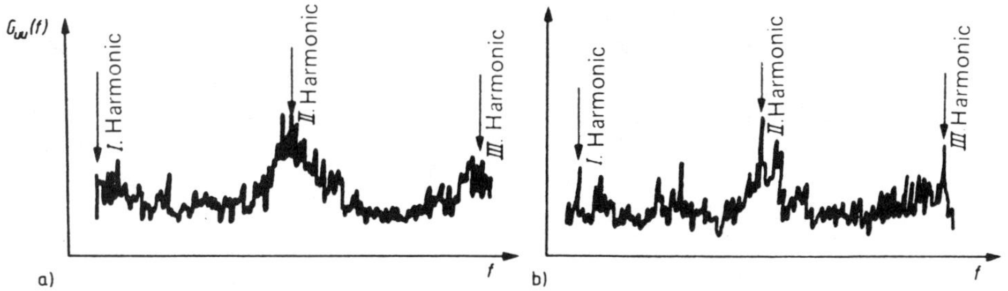

Fig. 3.36 — Vibration spectrum of a gear body: before (a) and after (b) replacement of gearwheels with frictional erosion [79]

The diagnostic application of cepstrum is useful for machines in which the revelation or disappearance of periodicity has a diagnostic significance (transmission gears, bearings and fluid-flow turbomachines, etc., etc.). It refers also to the subtle periodicities revealed in the amplitude–frequency modulation. Examples of such applications are shown in Fig. 3.36 and 3.37 which refer to the diagnostics of industrial transmission gears and taken from the work of Randall [79]. As we see, the frictional wear of the gearwheels has been revealed in the cepstrum rotation line 85 Hz which almost disappeared after wheel replacement. However, the diagnostic interpretation of cepstrum is not always so simple; hence, there have been attempts at finding other techniques for the treatment of spectral power density in order to reveal changes in its periodic part.

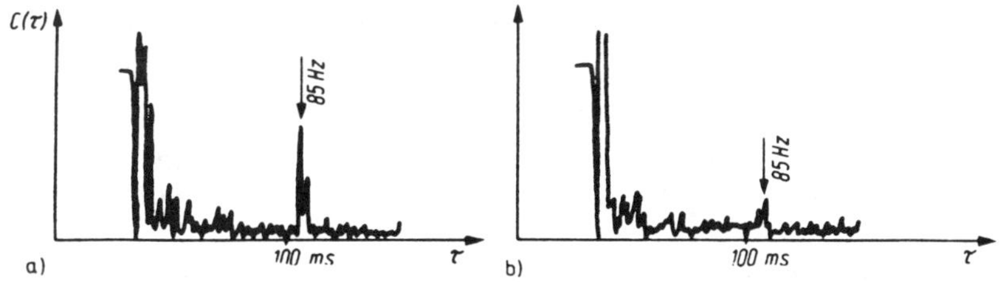

Fig. 3.37 — Cepstrum of gear: before (a) and after (b) replacement of gear wheels with a distinct change of amplitude to time τ corresponding to rotational frequency $f_0 = 85$ Hz [79]

Let us now discuss the *product spectrum* introduced in its original form by Schroeder in 1968 which in a modified form has been presented in the Author's monograph [7, Section 3.5.3]. The successive investigations on the utility of the definition given there have led to its modification giving finally the following forms of the product spectrum:

$$\text{И}(f_p) = \left[\prod_{n=1}^{N} G_{uu}(nf_p) \right]^{1/N} \quad \text{for } f_p \, \varepsilon \left(0, \frac{\Delta F}{3}\right) \tag{3.55}$$

As we can see form the definition, for the chosen frequency f_p, products are created here of its successive harmonics nf_p within the range of the existence of the original spectrum $f \, \varepsilon \, (0, \Delta F)$, and at the same time the range of the basic frequency f_p is limited to one-third of the interval of spectrum existence (-ΔF))/3. The final calculation of the N-th root from the successive products aims at the protection of the components with small amplitudes on the one hand, and on the other gives a physical sense to this value which is unusually significant and has not been preserved in the algorithm of cepstrum. As an example of such treatment of spectrum let us consider Fig. 3.38, presenting the spectrum of vibration acceleration of bearing in the fan BAB 106, its transformation into a product spectrum (Fig. 3.39(a)) and cepstrum (Fig. 3.39(b)). As we can see from the comparison of

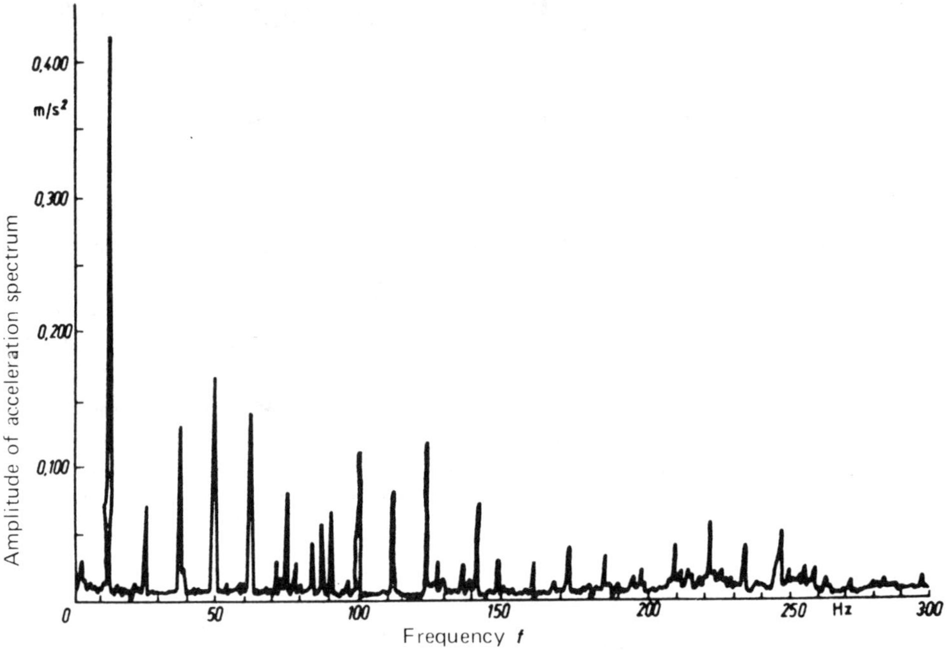

Fig. 3.38 — Vibration acceleration spectrum of fan BAB 106 bearing with visible harmonics of rotational frequency $f_0 = 12.3$ Hz [80]

Figs 3.39(a) and 3.39(b), they each contain the same information but, physically, it is more readable in the product spectrum $\text{И}(f_p)$ [80].

When we look at Equation (3.55), then instead of the product operation we can take into consideration the summation and change the square root by dividing through by the number of the sum elements. In this way we will arrive at the definition of a *polyharmonic spectrum*:

$$P(f_p) = \frac{1}{N} \sum_{n=1}^{N} G_{uu}(nf_p), \quad f_p \, \varepsilon \left(0, \frac{\Delta F}{3}\right) \tag{3.56}$$

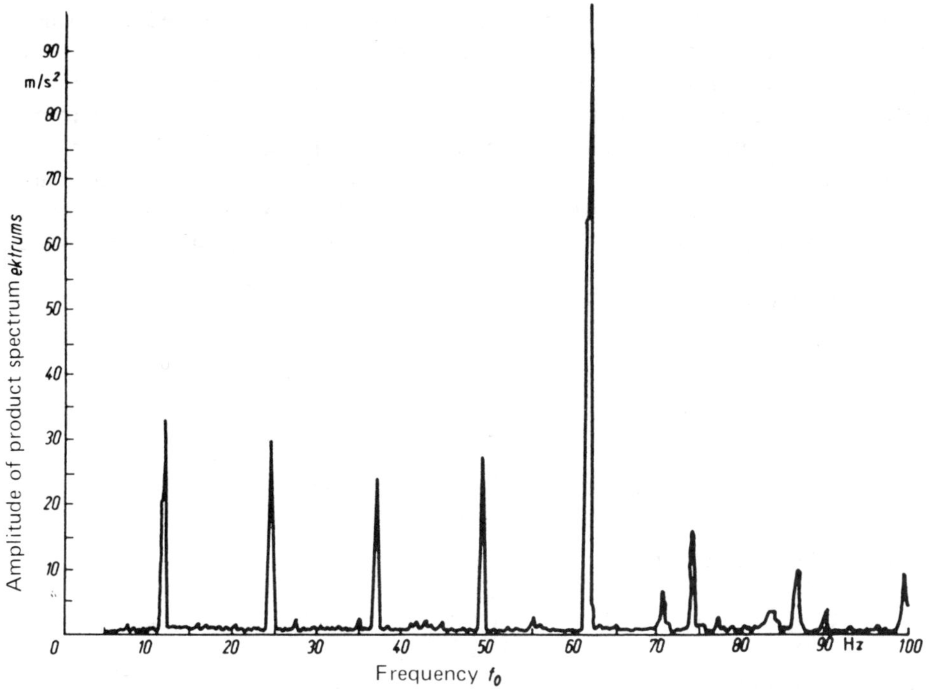

Fig. 3.39(a) — Product spectrum of vibration acceleration of fan BAB 106 bearing with visible harmonics of rotational frequency f_0 = 12.3 Hz [80]

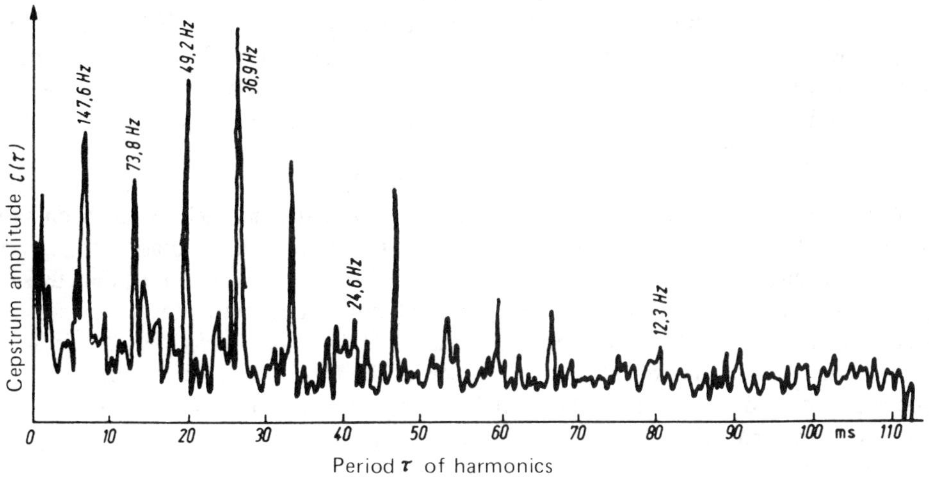

Fig. 3.39(b) — Cepstrum of vibration acceleration of fan BAB 106 bearing with visible harmonics of rotational frequency f_0 = 12.3 Hz [80]

which resembles in this character the action of a tunedperiodic filter from Fig. 3.20 and also the synchronous analysis proposed in the Author's monograph [7, Section 3.6.3].

An attentive analysis of the definition and results of the application of cepstrum $C(\tau)$, of the product spectrum И (f_p), and the polyharmonic $P(f_p)$ leads to the conclusion that

none of these procedures prevents the double consideration of the successive harmonics of the process. The Author [7] proposes in this connection the choice of f_p according to the prime numbers, but it creates essential difficulties in the division and covering of the investigated frequency band. Another solution to this dilemma is the elimination from the input spectrum of the already considered harmonics, so that each original spectrum line $G_{uu}(f)$ is taken into account only once. A diagram of the amplitudes of summed up harmonics in the function of basic frequency obtained in this way can be called the *spectrum of periodicity* $G_{T_p}(f_p)$. The definition of the operation leading to this spectrum is the following:
the successive line of periodicity spectrum:

$$G_{T_p}(f_p) = \sum_{n=1}^{N_p} G_{up}(nf_p) = G_{up}(f) * \frac{1}{T_p} \sum_{n=1}^{N_p} \delta(f - nf_p)$$

the current-filtered spectrum of the treated process:

$$G_{u_0}(f) = G_{u_{p-1}}(f) - G_{T_{p-1}}(f_{p-1}), \quad p = 1, \dots, \frac{1}{3}\,\mathrm{Ent}\left(\frac{\Delta F}{f_1}\right) \tag{3.57}$$

with initial assumptions:

$$G_{u_o}(f) = G_{u_u}(f), \quad G_{T_0}(f_0) = 0; \quad F_p \,\varepsilon \left(0, \frac{\Delta F}{3}\right)$$

where Ent $(\cdot)$ is the greatest integer, and ΔF is the frequency range of initial spectrum.

Therefore, each line component of periodicity spectrum $G_{T_p}(f_p)$ is the sum of amplitudes of the successive harmonics in the current spectrum $G_{up}(f)$ and the current spectrum of process $G_{up}(f)$ is created by the subtraction of the previous spectrum of line of periodicity $G_{uTp-1}(f_{p-1})$ from the previous current spectrum $G_{up-1}(f)$, and at same time $G_{up}(f) = G_{uu}(f)$ and $G_{T_0}(f_0) \equiv 0$.

This is the most recent conception and, so far, has not been experimentally tested. This concept is revealed to be very good and diagnostically oriented as has been shown recently by Mr R. Barczewski in his PhD Thesis. Comparing the formation of cepstrum, product spectrum and polyharmonic spectrum, we can say in summary that these are processes of the parallel filtration type. It means that in the formation of these spectra the whole original signal participates and, hence, the effects follow of sub- and superharmonics and the repeated consideration of the same lines. On the other hand, the above proposed operation carried out on the spectrum and, in consequence, on the signal $u(t)$, can be compared to the serial filtration as in Fig. 3.40, where the preceding filter cuts off from the next one its signal portion.

From the point of view of numerical filters, the proposed operation can be called a polyharmonic recurrent filtration since the filtration operation exerts a feedback on the form of the initial spectrum to the next operation $G_{up}(f)$, where:

$$p = 1, \dots, \frac{1}{3}\,\mathrm{Ent}\left(\frac{\Delta F}{f_1}\right).$$

Finally, it should be mentioned that the upper limit of the basic frequency:

$$p = \frac{1}{e} \operatorname{Ent}\left(\frac{\Delta F}{f_1}\right)$$

has been accepted by analogy to the product spectrum.

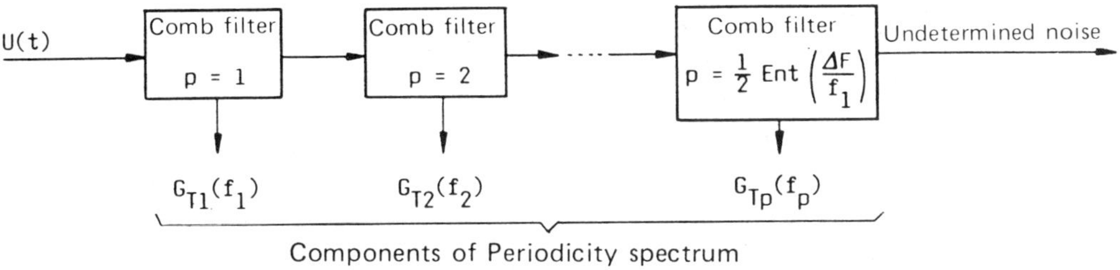

Fig. 3.40 — Periodicity spectrum in analogue comb filtration with a serial system of filters

3.5.6 Spectral sensitivity of vibroacoustical processes

We have already defined three types of measurement of vibro acoustical processes: (a) the dimensional amplitude measures according to Equation (3.48); (b) the dimensionless product measures according to Equation (3.49); and (c) frequency measures according to Equation (3.51). When we know the spectral density of the process power (e.g. vibration acceleration) then part of these measures can be calculated directly from spectrum $G_{aa}(f)$. We are using here the root-mean-square values of acceleration $\tilde{a}$, velocity $\tilde{v}$ displacement $\tilde{x}$: Rice's frequency : acceleration f_a, velocity f_v, displacement f_x ; and the index of the harmonicity of acceleration H_a, velocity H_v and displacement H_x . Generally, these measures have the right to be called the spectral measures, since they can be created from the spectrum of the original process, and the original process in vibrations is the spectrum of acceleration. The common origin of these measures permits us to investigate their common features, and primarily the sensitivity to changes in the character of spectrum, and in this way they permit us to select measures with the greatest sensitivity. This problem was the subject of investigations of a doctor's dissertation [81] and a further publication [82].

Let us consider Fig. 3.41 which presents the idea of spectrum synthesis of single lines with the amplitude G_k , half-width Δf_k and mid-frequency f_k. In the light of this idea, each measure ζ_j depends on these parameters creating:

$$\zeta_j = \Phi_j\left(G_k; \Delta f_k, f_k\right), \quad k = 1, \ldots, N \tag{3.58}$$

and sensitivities to changes of these parameters are defined as the logarithmic derivatives:

$$W_{\zeta_i}^{G_k} = \frac{\partial \ln \zeta_i}{\partial \ln G_k}, \quad W_{\zeta_i}^{f_k} = \frac{\partial \ln \zeta_i}{\partial \ln f_k}, \quad W_i^{f_k} = \frac{\partial \ln \zeta_i}{\partial \ln f_k} \tag{3.59}$$

For definite spectra of vibrations, the above analysis of sensitivity gives quantitative guidelines regarding the diagnostic orientations of the particular measures. An example of the results of such analysis is shown in Fig 3.42, which indicates that spectral low-frequency changes will be most perceptible by the measure of the root-mean-square

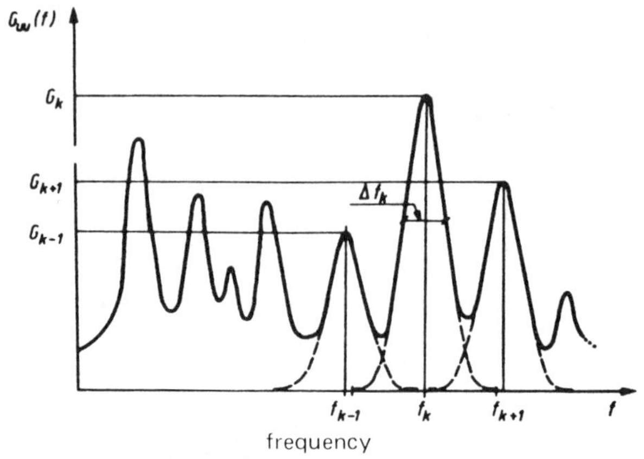

Fig. 3.41 — Idea of the synthesis of acceleration spectrum from elementary lines with parameters $G_k, f_k, \Delta f_k$ [82]

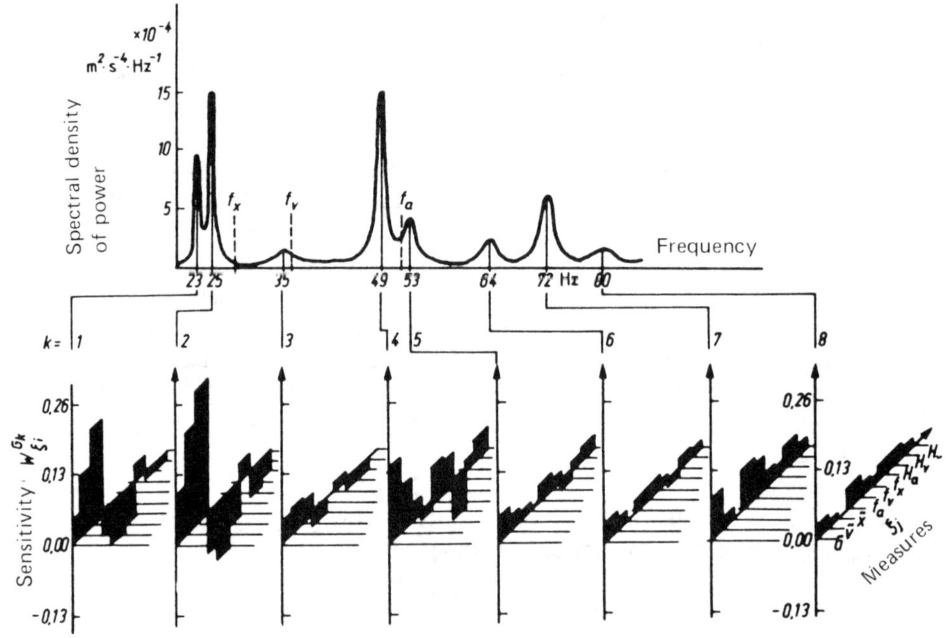

Fig. 3.42 — Calculation example of the sensitivity of spectral measures to the change of the height of spectrum lines of electric motor vibrations [82]

displacement and in the fields of average frequencies by Rice's frequencies f_a, f_v.

The analysis of sensitivity permits us also to draw a number of general conclusions independent of the form of the spectrum of the discussed process. A summary of the spectral orientation of the particular measures is presented in Table 3.3 where f_{min} and f_{max} denote the minimum and maximum frequency of the spectrum, while f_x, f_v, f_a are Rice's frequencies of displacement, velocity and acceleration of this spectrum.

Table 3.3 Comparison of the sensitivity of spectral measures in the particular intervals of spectral frequency [82]

Oriented change of spectrum parameter	The most sensitive measures in the interval of spectral frequency of the power $G_{aa}(f)$			
$G_{aa}(f: G_k, \Delta f_k), f_k$ (failure symptom)	$f \varepsilon (f_{\min}, f_x)$	$f \varepsilon (f_x, f_v)$	$f \varepsilon (f_v, f_a)$	$f \varepsilon (f_a, f_{\max})$
– change of G_k, e.g. intensity of failure	$\tilde{v}, \tilde{x}$ f_v, f_x	f_x	From the second class of sensitivity $\tilde{a}$	f_a
or				
– change of Δf_k, e.g. change of dissipative properties of the system	H_a, H_v, H_x	H_a	H_a	H_a
– change of f_k, e.g. change of elasticicity or mass of the system	$\tilde{v}, \tilde{x}$ f_v, f_x H_a, H_v, H_x	$\tilde{v}$ f_v $H_a, H_v,$	f_a $H_a, H_v,$	f_a H_a

It follows from Table 3.3 that if we look for changes of high frequencies $f > f_a$, then the best alternative is just the measurement of f_a or H_a. On the other hand, for the low-frequency band, the wealth of possibilities is significantly greater, while the most frequently used measure with high sensitivity is H_a the index of the acceleration harmonicity or the original process. It means that in investigating the relative width of the band of the original process we can find the greatest number of changes in the vibroacoustical process.

3.6 SUMMARY

In this chapter, we have passed from the vibroacoustical process through the vibroacoustical signal and its measures leading to the final evaluation of the technical condition. We have presented models of diagnostic signals (Section 3.2), models of the generation of these processes and signals (Section 3.3) and the possible ways of a preliminary filtration of the useful signals (Section 3.4). Finally, we have presented the premises for the selection of diagnostic measures for vibroacoustical signals, i.e. measures commonly used in diagnostics as well as some newly formulated ones. In the light of the above we have already sufficient knowledge to define the condition of a machine by selecting: (a) the proper vibroacoustical processes; (b) the point of measurement; and (c) the vibroacoustical measure of signal oriented towards the failure. Now we have to clarify the transfer from a quantitative value of the signal measure (i.e. vibroacoustical symptom) to the technical condition. This is the diagnostic inference, to be discussed in the following chapter.

4

Diagnostic inference

4.1 INTRODUCTION

The cause-and-effect chain in vibroacoustical diagnostics (failure → vibroacoustical process → vibroacoustical signal → vibroacoustical symptom of condition → CONDITION ASSESSMENT) was introduced in the preceding chapter. From our earlier considerations we know how to arrive at the the vibroacoustical symptom of the technical condition; therefore, the next stage is to carry out, on this basis, the *diagnosis of the mechanical condition*. Let us recall that under the notion of complete diagnosis we understand the evaluation of the actual condition (diagnosis), the determination of the causes of this condition (origin) and determination of the causes of the time horizon of the future change of this condition (prognosis). Let us recall also that the notion of technical condition is widely understood as: (a) the dynamic condition of the prototype of the machine during its construction; (b) the condition of the quality of the element or machine assembly in the process of its manufacturing; and (c) the utilization condition of the element or machine assembly during its use. At the same time, it is obvious that the stage of prognosis within the complete diagnosis can be applied only in the use of diagnostics, where the development of the failure and the degradation of the technical condition take place together,

The decision on the technical condition is possible only when we know the diagnostic model of the mechanical object i.e when we know the overt or covert assignment of the vibroacoustical SYMPTOM → CONDITION. The use of this relation requires the removal of incidental disturbances in the observation of vibroacoustical symptoms of the condition and the determination of the condition class at least according to a two-condition division. Finally in the use of diagnostics, the ability to determine the future changes of this condition is also necessary. This problem will be clarified in detail in this chapter.

4.2 DIAGNOSTIC MODELS AND DECISIONS CONCERNING THE CONDITION OF MECHANICAL OBJECTS

In Section 3.2.1, we introduced the notion of the curve of failure life $F(\theta)$, which reflects in a quantitative way the advancement of failure or the real utilization condition of the object in categories of one failure. It rarely occurs that our numerical manifestation of the condition refers only to one failure. This happens for two reasons: (a) the region of failure in the object is generally multidimensional; and (b) the real life curves are generally linear or nonlinear depending on each other. Therefore, if we have an n-dimensional space of failure, then theoretically we could describe its condition by a set of life curve values in the form of a condition vector:

$$\{\mathbf{F}(\theta)\} = \text{col}\left\{\mathbf{F}_1(\theta), \mathbf{F}_2(\theta), \ldots \mathbf{F}_n(\theta)\right\}, \text{ for } \theta = \theta_1 \theta_2 \ldots \tag{4.1}$$

However, in practice, this condition must be described by another set of numbers:

$$\{\mathbf{X}(\theta)\} = \text{col}\left\{\mathbf{X}_1(\theta), \mathbf{X}_2(\theta), \ldots \mathbf{X}_n(\theta)\right\}^T \tag{4.2}$$

At the same time generally $X_i(\theta) = \sum_j^n a_{ij} F_j(\theta)$, while col $(\cdot)$ denotes the column vector, θ is the time of the objects use (when we deal with diagnostics of quality, then time θ will pass to the initial number of the object) a_{ij} some factors of the weight of the linear development of the relation between the *measurable characteristics of condition* X_i *(*θ*) and the original characteristics of condition* Fj *(*θ*).*

In the vibroacoustical diagnostics of machines, it is necessary to know for the determination of the condition at least n vibroacoustical symptoms: $S_k(\theta)$; $k = 1, \ldots, m \geq n$ creating the vector of symptoms:

$$\{\mathbf{S}(\theta)\} = \text{col}\left\{\mathbf{S}_1(\theta), \ldots, \mathbf{S}_m(\theta)\right\}^T \tag{4.3}$$

and we must know the relation between symptoms $S(\theta)$ and the condition characteristics $X(\theta)$ where it is not necessary to distinguish the vector from its components.

This relation can be overt, e.g. in the form of a functional relation found by the *regression* method. It can also be *covert* but known from the exemplary training set of good machines. Finally, this relation which is by nature a *cause-and-effect* one, can be observed in the presence of such a high level of disturbances that we are forced to apply the *probabilistic model* to its description. In each of the mentioned cases of the relation SYMPTOM → CONDITION there exist two methods of diagnostic inference as discussed below.

4.2.1 Diagnostic regression model

Information about the character and the form of the relation condition–signal can be obtained only by experiment. At the same time one can distinguish three types of diagnostic experiment: The first is an active experiment where we have at our disposal an optional possibility of changing the condition characteristics $X(\theta)$ and we observe in a definite place r of the object the vibroacoustical symptoms of its condition $S(r, \theta)$. The second type is an *active–passive experiment* in which there is no possibility of changing the value of the condition characteristic $X(\theta)$, but we possess information about the point values of these characteristics, e.g. in a new object $X(0)$ when setting it into operation, during stoppage for overhaul it is $X(\theta_{rem})$, or in post-overhaul staring $X(\theta_0)$. On the other hand, we observe in a continuous or discrete way the vibroacoustical symptoms of the condition $S(r, \theta)$. Finally, the third type is a *passive experiment* consisting of the observation of vibroacoustical symptoms of the object's condition without the quantitative information about the condition characteristics. We only possess information about the object's actual *condition class*, e.g. good or faulty condition.

Let us assume now that we have carried out an active experiment on the object, and its results elaborated by the statistic regression method will give us the following condition–signal relation which will be called the *diagnostic model* of the object [49, 83]:

$$\{\mathbf{S}(r,\theta)\} = \psi[r, X(\theta)] + \{\mathbf{N}_x(r,\theta)\} \cong A(r)\{\mathbf{X}(\theta) + \{\mathbf{N}_x(r,\theta)\} \quad (4.4)$$

where $\psi[\cdot]$ is the matrix operator of relationship, $\{\mathbf{N}_x(\cdot)\}$ is the vector of uncontrolled disturbances, and $A(r)$ is the matrix of the sensitivity of the linear approximation of operator $\psi[\cdot]$. A similar diagnostic model by with a different content will be obtained from a passive–active experiment.

If, instead of the unknown vector of condition characteristics $\{\mathbf{X}(\theta)\}$, we substitute the vector of the original *vibroacoustical symptoms* (interchangeable with the elementary failure) denoted by $\{\Lambda(\theta)\}$, then instead of Equation (4.4) in the field of condition vector we will have:

$$\{\mathbf{S}(r,\theta)\} = \psi[r, \mathbf{L}(\theta)] + \{N_0(r,\theta) \cong A(r)\{\Lambda(\theta)\} + \{\mathbf{N}_\theta(r,\theta)\} \quad (4.5)$$

We have now obtained a diagnostic model of a machine in the form of a set of life curves in which each of the components of symptom vector is a weighted sum of the original life curves $\Lambda_j(\theta)$ and a certain portion of disturbances:

$$S_i(r,\theta) = \sum_{j=1}^{n} A_{ij}(r)\Lambda_j(\theta) + N_{i\theta}(r,\theta) \quad (4.6)$$

where $A_{ij}(r)$ forms the elements of the known matrix of sensitivity $A(r)$, $\Lambda_j(\theta)$, denotes the original life curves depending on the curves of failure life $F_j(\theta)$. and $N_{i\theta}(r, \theta)$ is the component disturbing the i-th symptom. It is worth noting here that, with a skilful selection of the vector of symptoms $\{\mathbf{S}(r, \theta)\}$, matrix $A(r)$ can be diagonal, and then the life curves $\Lambda(\theta)$ will be identical with the failure life curves $F(\theta)$ (Section 6.3).

The diagnostic models in Equations (4.4) and (4.5) have the following common characteristics: (a) they are obtained as the result of regressive processing of the experimental data; (b) they consist of the active experiment for Equation (4.4) and of the passive experiment for Equation (4.5). So, we can call them the *regressive models*, while in vibroacoustical diagnostics a significantly more common model is the regression of symptom S (or the vector of symptoms) in the domain of lifetime θ. In both diagnostic variants of the regressive model the diagnostic inference consists of the comparison of the current value of symptom S with its limit value S_l With a two-class division of the technical condition into the good/faulty classes, the inference for one symptom is the following:

$$\begin{aligned} S < S_l &\rightarrow \text{Good Condition} \\ S > S_l &\rightarrow \text{Faulty Condition} \end{aligned} \quad (4.7)$$

Such reasoning should be applied for each component of the vector of symptoms, but to obtain a total measure of the condition of the object with n-dimensional space of failures is not simple. It depends on the quantitative and qualitative influence of partial failures on the total condition of the object. It may be an excluding influence on the type 'everything or nothing', or a gradual deterioration of the total condition. In the first case we can apply Boolean algebra and in the second the cumulation measures (e.g. the product measures) comprising all failure symptoms in one generalized symptom. Such an approach is very attractive since, for the evaluation of the condition, we can apply the reasoning based on the excedence limit value according to the scheme in Equation (4.7). The determination of these limit values is not simple and, even in the model according to the active experiment, must be based on statistical investigations owing to the spread of the properties which have been discussed in the Introduction to this book. So, in the best arrangement of experiments, adequate distribution densities of the following values are available: (a) the initial value of the symptom for new object condition $p(S_0)$; (b) the limit value of symptom $p(S_a)$; (c) the limit value of condition characteristics $p(X_a)$; as well as (d) the distribution density of operation time until failure $p(\theta_a)$. The above-mentioned distributions, regression models and methods of diagnostics inference are presented in Figs 4.1(a) and 4.1(b), being true for a machine with one failure or with an n-dimensional area of failure using the cumulative measures.

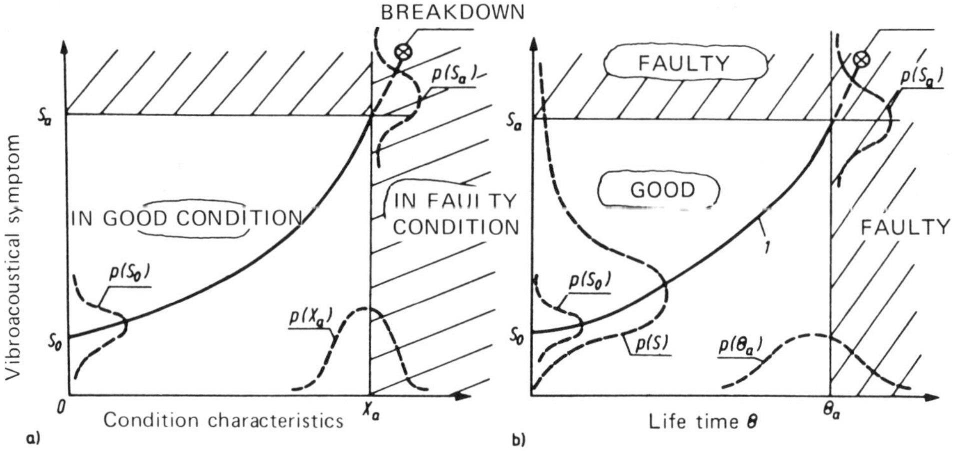

Fig. 4.1 — Regressive diagnostic model of a machine in the domain of condition characteristics (a), and lifetime (b) with the corresponding limit values X_a, θ_a, S_a, S_0 following from the distribution of these parameters

As we can see in Fig, 4.1, the determination of the limit value S_a of vibroacoustical symptom has an indirect character for the regression in relation to condition characteristics, since the decisive characteristic in the active experiment is the condition characteristic X. It is different for the regression in the domain of lifetime. Here, independently of the accepted utilization philosophy, the value of vibroacoustical symptom S can be a decisive variable if we minimize the level of unnecessary overhauls, or an auxiliary variable in the system of planned preventive overhauls. Hence, the difference between the form of the density condition characteristics distribution $p(X_a)$ and sometimes the lifetime $p(\theta_a)$, since the average time of operation until failure θ_a depends not only on X_a but also on other factors. Therefore, the form $p(\theta_a)$ is more flattened and the utilization based on a planned preventive overhaul brings fewer benefits than one based on vibroacoustical diagnostics.

4.2.2 Nonexplicit model of image type

The cause-and-effect connection between the vibroacoustical symptom and condition characteristic X or operation time θ, is not always known in the analytical sense, in spite of the fact that it always exists and constitutes the basis of diagnostics. In the diagnostics of objects with multidimensional areas of failures, the assignments of symptom → failure, or symptom → condition, symptom → lifetime, are not always known. In such cases we know, for example, that for the given condition the vector of vibroacoustical symptoms $\{S\} = \mathrm{col}\{S_i, \ldots, S_n\}$ should have components with definite or approximate values, but it is not clear why it should have such values (unknown assignment regarding the characteristics of condition X), and when it will take place (unknown assignment in relation to lifetime θ).

Each mechanical object is described in our case by vector $\{S\}$ in n-dimensional area of vibroacoustical symptoms, and we can say that the set of coordinates of this object $S_1, \ldots, S_n$ creates the *pattern or image* in the n-dimensional area. Thus, the recognition of condition X representing this object can be treated as the *vibroacoustical recognition of the condition image* given by $\{S\}$ when we know how to do it! It is possible to learn how to recognize when we have the vectors of symptoms $\{S\}_p$ for a group of objects $p = 1, \ldots, k$ with a known condition of serviceability. Then we speak about the known training sets:

$$\begin{aligned} \{S\}_p &\equiv \{S^z\} \rightarrow \text{GOOD CONDITION}, \qquad p = 1, \ldots, k \\ \{S\}_p &\equiv \{S^n\} \rightarrow \text{FAULTY CONDITION}, \qquad p = k+1, \ldots, r \end{aligned} \tag{4.8}$$

where $\{S^z\}$ denotes the area of good condition, and $\{S^n\}$ is the area of faulty condition in the n-dimensional area of vibroacoustical symptoms.

The recognition of the GOOD CONDITION of another object with image $\{S\}_i$ will now follow according to the relation of the known training sets:

$$\{S\}_i\ \varepsilon \cdot \{S^z\} \rightarrow \text{GOOD CONDITION}$$

or (4.9)

$$\{S\}_i\ \varepsilon \cdot \{S^n\} \rightarrow \text{FAULTY CONDITION}$$

Now, depending on the way of the determination of the membership of the training sets, we will have different methods of pattern recognition (see, for example, Tau and Gonzales [84]) and in the application to diagnostics we see the monograph by Birger [12] and the dissertation by Cholewa[85]. Generally, the recognition methods can be divided into two groups: (a) the distance methods which build the membership function on different measures of distances (of Euclid, Haming, etc.) of the recognized image from the training sets; and (b) the division of the area of symptoms into a subarea of the GOOD AND FAULTY CONDITIONS with the help of decision hypersurface $\Phi(\{S\}) = 0$, and it has a particular application in the two-class division of condition. Since the equation of such a surface is not always easy to find, we have a whole range of methods of direct determination of this surface. The effect of the action of hypersurface itself is best illustrated by the recognition of cases with one-dimensional and two-dimensional symptoms as in Figs 4.2(a) and 4.2(b).

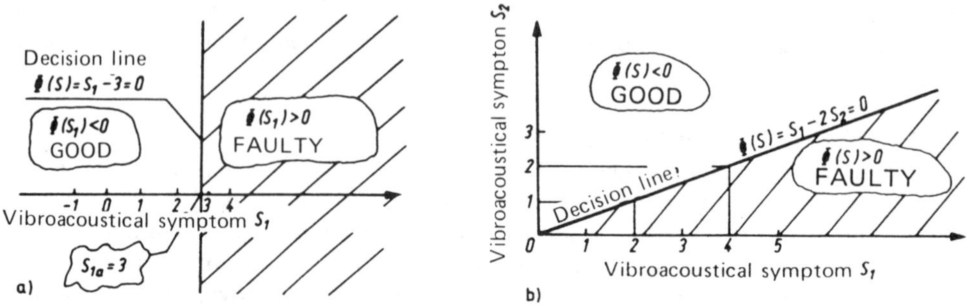

Fig. 4.2 — Recognition of vibroacoustical pattern of the object's condition in one-dimensional (a) and two-dimensional (b) patterns

The analysis of both figures shows analogies to the regressive models. The determination of decision lines in the one-dimensional case is equal to the determination of the limit value for symptoms S_1 equal to S_{il} . The analogy of the two-dimensional case is very subtle, but nevertheless is obvious.

It follows, therefore, that with a preset decision hypersurface, the rule of classification (recognition) of the condition should be the following:

$$\operatorname{sgn}\Phi(\{S\}_i = \begin{cases} +1, \{S\}_i \ \varepsilon \ \{S^n\} \rightarrow \text{Faulty Condition} \\ -1, \{S\}_i \ \varepsilon \ \{S^z\} \rightarrow \text{Good Condition} \end{cases} \tag{4.10}$$

where the function of sign sgn $(\alpha) = +1$ for $\alpha \geq 0$ and sgn $\alpha = -1$ for $\alpha < 0$.

Having explained the recognition of the condition in the multidimensional area of vibroacoustical symptoms, let us proceed to the determination of the hyperplane $\Phi\ \{S\} = 0$. In many cases of diagnostics and the two-condition evaluation of the condition, the dividing hypersurface can have the form of a single or broken hyperplane. In the first case, the decision function being an equation of hyperplane has the form:

$$\Phi(\{S\}) = \{\lambda\}\{S\} = \sum_{i=1}^{k} \lambda_j S_j = 0 \tag{4.11}$$

where $\{\lambda\}$ is an unknown row vector of weights such that the scalar product of both vectors is zero.

The latter relation and decision rule in Equation (4.10) presents the operation method of the so-called linear classifier and its whole problem is reduced to the calculation of the components λ_i of the vector of weights $\{\lambda\}$ on the basis of vibroacoustical images of objects with known conditions. The procedure for finding the vector $\{\lambda\}$ bears the name of the learning procedure and, according to Birger [12] and Tau [84] can be carried out as a sequence of iterations on known patterns $\{S\}_p$. Taking the first pattern $\{S\}_1$ we can write the first approximation of vector $\{\lambda\}$ in the form:

$$\{\lambda\}_1 = \{S\}_1 \text{ if } \{S\}_1 \ \varepsilon \ \{S^n\} \qquad \text{Faulty Condition}$$

or (4.12)

$$\{\lambda\}_1 = \{S\}_1 \text{ if } \{S\}_1 \ \varepsilon \ \{S^z\} \qquad \text{Good Condition}$$

The decision plane for this approximation is defined by the equation $\Phi_1\ \{(S)\} = \{\lambda\}\ \{S\} = 0$, while generally for the n-th approximation it is $\Phi_n\ (\{S\}) = \{\lambda\}_n\ \{S\}$

$$\{\lambda\}_n = \{\lambda\}_{n-1} + 1/2\,[\mathrm{sgn}\,\phi\,(\{S\}_n) - \mathrm{sgn}\,\phi_{n-1}\,(\{S\}_n)]\,\{S\}_n \tag{4.13}$$

where $\{S\}_n$ is the successive training patternwith known condition and known sign of decision function ϕ $(\{S\}_n)$ and $\phi_{n\ -1}$ $(\{S\}_n)$ are the values of the n th and 1 th approximation of this function for pattern $\{S\}_n$.

It can be shown that the sequence of such iterations leads within limits to the calculation of vector $\{\lambda\}$ and thereby to the finding of the linear form of decision function ϕ $\{S\}$. In practical calculations it is convenient to extend the dimension of vector $\{S\}$ by adding one as the first component. Then vector $\{\lambda\}$ can also comprise the cases of hyperplanes not passing through the centre of the system of coordinates. Among other methods of recognition with a simple logical structure it is worth mentioning the method of the closest neighbour and the method of hyperzones [86]. In the first case, the decision function is a broken hyperplane, while in the second it consists of hyperzone segments. The realization apparatus of both methods and the realization of linear classifier is very simple and can be carried out with the help of diodes and resistors; in a still simpler way it can be done using standard minicomputer programs.

For the correct determination of the form of decision function according to these methods we need a training set of the size of several dozen objects. After the determination of the learning process, we can proceed to the automatic recognition of the machine condition on the basis of the measured set of vibroacoustical symptoms $\{S\}$ being the vibroacoustical image of the object's condition.

We have considered the simplest version of the covert diagnostic model of the pattern type limiting our information to the linear and determined methods of inference. The recognition of patterns is a dynamically developing discipline and its many achievements can be adapted for diagnostic purposes (Chapter 6).

4.2.3 Probabilistic diagnostic model of a mechanical object

The above constitutive relationships of diagnostics (i.e. the regressive models in Equations (4.4) and (4.5), and the model of pattern type in Equation (4.8) can be obtained at a low level of disturbance in the vibroacoustical symptoms observation (N_x and N_θ are components in the above models). With high-level disturbances, our constitutive relationships pass from the deterministic to the probabilistic ones, allowing us to make the final decision about the condition in the categories of the probabilities of the good/faulty condition instead of the previous certainty about the existing condition. Seemingly, the situation looks dramatic, but it is not so bad, since in the previous models there always existed certain levels of disturbance giving a diagnostic reliability or confidence less than 100%. However, previously we have accepted that without undertaking any special steps. But now, with a further decrease in confidence(e.g. 80%) we are forced to look in a different way at the problem of our diagnostic knowledge and ignorance, admitting that for the given symptom S the object may be either in a good or in a faulty condition. Thus, our diagnostic model will consist in the assignment of *conditional probabilities* for both mutually excluding events. In the categories of the density of probabilities, our *diagnostic probabilistic model* for one symptom S can be written in the following way:

$p\,(S\,/\,z)$ = density of the probability of GOOD CONDITION with known symptom S
$p\,(S\,/\,n)$ = density of the probability of FAULTY CONDITION with known symptom S (4.14)

This model will be more obvious when we take the probability of both event for a definite value S :

$$P(S/z) = \int_{-}^{S} p(S_k/z)\,\mathrm{d}S_k, \quad P(S/n) = \int_{-}^{S} p(S_k/n)\,\mathrm{d}S_k \qquad (4.15)$$

with the obvious condition of a *complete probability of the occurrence of symptom* S:

$$P(S) = P(S/z)\,P(z) + P(S/n)\,P(\mathrm{n}) \qquad (4.16)$$

where $P(z)$ and $P(n)$ are the supplementary $P(z) + P(n) = 1$ of the probability of good and faulty conditions.

The above probabilistic model for the interpretation for one symptom is also true for the vector of symptoms {S}; it is necessary only to make an adequate transition in Equation (4.14) and to understand the integral (Equation (4.15)) as a multidimensional one.

In the recognition of the condition according to the probabilistic model (Equation (4.14)) we can distinguish two groups of methods. The first is based on the decision about the condition of the transgression of the set threshold of recognition P_x, as it is in Bayes's method, or in Wald's, sequential method. The second group bases the decision concerning the condition of the object on the transgression of the limit values of the vector of signal S_l, as in the regressive model. However, here the values of S_l are determined by the theory of statistical decision with the help of numerous detailed techniques.

The transgression of the set threshold of the accuracy of diagnosis [12] is an inference rule in Bayes's method and Wald's sequential method.

In Bayes's method, we distinguish a priori the knowledge we possess before the trial to evaluate the condition, and the a posteriori knowledge obtained as a result of the application of the method of statistical inference. So, for the application of the Bayes's method of inference it is necessary to possess the valuation of the probability of the GOOD CONDITION $P(z)$ and the FAULTY CONDITION $P(n)$ in the given group of machines. Since the probability of the GOOD CONDITION of a machine group in operation is equal to the *index of availability* $P(z)$ [87], it is enough to know this index since $P(z) + P(n) = 1$. The successive evaluation of probabilities which we should know, is the conditional probability of occurrence of value S of the symptom in GOOD CONDITION $P(S/z)$ and in FAULTY CONDITION $P(S/n)$.

As we know, the unconditioned occurrence of condition $X = \{\mathbf{z}, \mathbf{n}\}$ (good or faulty condition) with the values of symptom S (i.e. $P(XS)$) can be calculated from the formula [77]:

$$P(XS) = P(X/S)\,P(S) = P(S/X)\,P(X) \qquad (4.17)$$

and from the two latter alternatives there follows Bayes's theorem of the probability of the occurrence of condition X with the measured symptom S:

$$P(X/S) = P(X)\frac{P(S/X)}{P(S)} = \frac{P(X)\,P(S/X)}{P(z)\,P(S/z) + P(n)\,P(S/n)} \qquad (4.18)$$

where $X = \{\mathbf{z}; \mathbf{n}\}$ denotes the alternative GOOD CONDITION '*z*' or the FAULTY CONDITION '*n*', while $P(S)$ is a complete probability of the occurrence of symptom S substituted according to Equation (4.16). Thus, having the preliminary information about $P(X)$ and defining $P(S/X)$ from the measurement of the value of symptom S, we can calculate the probabilities of the GOOD CONDITION $P(z/S)$ or FAULTY CONDITION $P(n/S)$ with a given S. If the probabilities calculated in this way exceed the threshold of recognition i.e.:

$$P(X/S) > P_x, \text{ for } X = \{\mathbf{z}, \mathbf{n}\}$$

then the value s will determine the condition with the greatest probability:

$$P(z/S) \gtrless P(n/S) \tag{4.19}$$

or the condition for which

$$\frac{P(z/S)}{P(n/S)} \gtrless 1 \tag{4.20}$$

is greater than 1.

As mentioned before, S can be a single symptom or a vector $\{S\}$. In the latter case, according to Equation (4.18), we have to calculate the conditional probabilities $P(\{S\}/X)$ which can easily be estimated, assuming the independence of the vector components $\{S\}$ = col $\{S_1, \ldots, \{S_n\}$ which, according to Birger [12], is not the source of large errors. Then we have:

$$P(\{S\}/X) = \prod_{j=1}^{n} P(S_j/X) \tag{4.21}$$

and we can use Equation (4.18) again.

Summing up *Bayes's method of* diagnostic *inference* we can conclude that, after the transgression of the recognition threshold in Equation (4.19), we accept for the object the condition with a maximum probability according to Equation (4.20). The accuracy of inference can be significantly increased if our knowledge a priori about the object differs significantly from the knowledge of the uniform distribution $P(z) = P(n) = 1/2$.

If the signal vector has components with different information weight, then the measurement of all components may sometimes prove futile. This is anticipated by the application of a sequential analysis in diagnostic decision-taking. This method is based on the value of the quotient of opposite events (good/faulty), and as our criterion we use the product of such expressions calculated for the successive components of vector $S_1, \ldots, S_n$ with a decreasing information significance:

$$L_t = \prod_{j=1}^{t \le n} \frac{P(S_j/z)}{P(S_j/n)} \tag{4.22}$$

The decision rule of Wald's sequential method is now as follows (Fig. 4.3):

$$L_t \cong \begin{cases} > A, \; \{S\} \in \{S^z\} \rightarrow \text{FAULTY CONDITION} \\ < B, \; \{S\} \in \{S^n\} \rightarrow \text{GOOD CONDITION} \end{cases} \tag{4.23}$$

$B < L_t < A \Rightarrow$ take into account the successive component t of vector $\{S\}$

The limits A and B of a reliable diagnosis are calculated from special formulae [12, 88] taking into account the admissible probability of the unnecessary overhaul and the unperceived breakdown as a function of the number of considered components of the vector of symptoms as in Fig. 4.3. This method therefore has the merit that the decision about

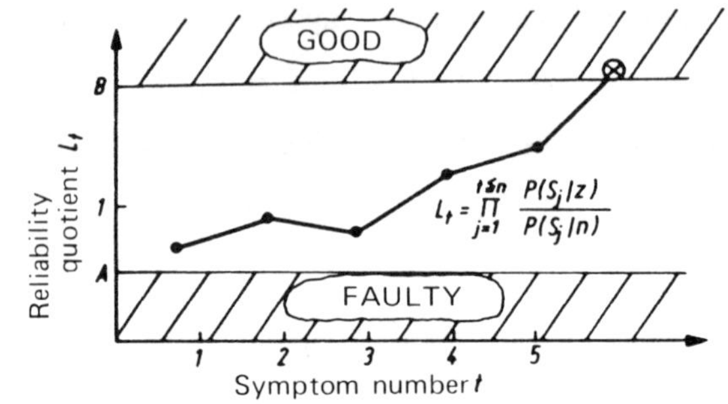

Fig. 4.3 — Sequential method of diagnostic inference and A, B limits for reliability quotient

the object's condition can be taken only on the basis of several most informative components of the signal vector (the sixth component has proven sufficient for decision-taking of Fig. 4.3). This gives the possibility of minimizing the set of checkouts and the costs of diagnosis, particularly if the components of the vector of symptoms are interdependent.

Diagnosis as a statistical decision

The probabilistic diagnostic model of the object formulated by Equations (4.14) and (4.16) allows another method of inference with the help of the limit value of symptom S_l. The limit value can be obtained with the help of the theory of *statistical decision* after the definition of the total *risk of diagnostic decision*. For a clear explanation, let us use here, instead of the signal vector, its one component $S_i = S$. For this component the probabilistic model in Equation (4.14) can be illustrated graphically in the form of the density of symptom probabilities in the GOOD CONDITION $p(S/z)$ and FAULTY CONDITION $p(S/n)$ as in Fig. 4.4.

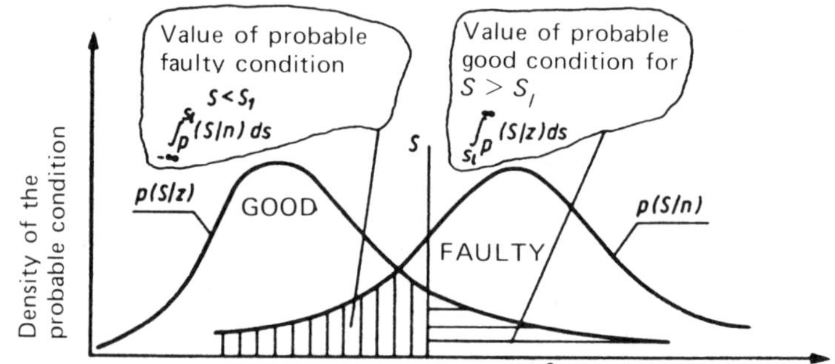

Fig. 4.4 — Probabilistic model of the diagnosis of a mechanical object as an introduction to the determination of the limit value S_l with the help of the theory of statistical decisions [89]

For the sake of accuracy, it must be stated that the density of the FAULTY CONDITION $p(S/n)$ in vibroacoustical diagnostics can have only a visual significance if the criterion of FAULTY CONDITION is an objective reason such as breakdown immobilizing the machine (since after breakdown we cannot measure the vibrations of the machine). On the other hand, if we define the FAULTY CONDITION subjectively (e.g. as a transgression of the standard vibration value) then further considerations of both distribution densities have a full logical sense. To the set of our assumptions let us assume that we know the probability of GOOD CONDITION (availability) $P(z)$ of the analyses group of objects and the corresponding probability of FAULTY CONDITION $P(n)$. Then, having this knowledge, on the basis of the model from Fig. 4.4, we can define the following events.

probability of the correct estimation of GOOD CONDITION

$$P(z) \cdot \int_{-}^{S_l} p(S/z)\, dS = P(z)\, P\left(S < S_l / z\right)$$

probability of a false alarm or *unneeded overhaul* following a faulty estimation of the condition

$$P(z) \cdot \int_{S_l} p(S/z)\, dS = P(z)\, P\left(S > S_l / z\right)$$

probability that the FAULTY CONDITION—breakdown will not be detected (4.24)

$$P(n) \cdot \int_{-}^{S_l} p(S/n)\, dS = P(n)\, P\left(S < S_l / n\right)$$

probability of a correct evaluation of the FAULTY CONDITION (breakdown)

$$P(n) \cdot \int_{S_l} p(S/n)\, dS = P(n)\, P\left(S > S_l / n\right)$$

Assigning now to each event the corresponding generalized costs C_{ij} in the form of a penalty for a faulty '*f*' estimation of the condition and a reward for a correct '*t*' estimation of the GOOD CONDITION '*z*' and the FAULTY CONDITION '*n*', we can formulate after Birger [12] the global estimation of the risk R of a diagnostic decision in the form:

$$\begin{aligned} R = C_{tz} \cdot P(z) \cdot \int_{S_l} p(S/z)\, dS + C_{fz} \cdot P(z) \cdot \int_{S_l} p(S/z)\, dS \, + \\ C_{fn} \cdot P(n) \cdot \int_{-}^{S_l} p(S/n)\, dS + C_{tn} \cdot P(n) \cdot \int_{S_l} p(S/n)\, dS \end{aligned} \qquad (4.25)$$

As we can see from the above, it is a complicated functional equation, generally in a covert form, for the determination of the limit value of symptom S_l. According to Birger [12], there are six methods for the determination of S_l minimizing the risk ($R = R_{min}$) differing by the accounted number of factors in the estimation of the risk in Equation (4.25) as well as by the degree of a priori knowledge of the probabilities $P(z)$ and $P(n)$. However, in each case we can write the qualitative formula for the determination of S_l from Equation (4.25) in the form:

$$S_l = S\left(R = R_{\min}\right) \qquad (4.26)$$

Without the calculation details of these methods and referring the interested readers to literature (see, for example, Birger [12] and de Groot [90]) we will mention and characterize only the most important calculation techniques. The most general method is that of the minimum risk, where we take into consideration all components of risk R in Equation (4.25). In the methods of the minimum number of erroneous decisions, we take into consideration only the middle members of Equation (4.25) with $C_{fz}, C_{fn} \neq 0$. In turn, in the 'minimax' method, we do not know the a priori probabilities $P(z)$ and $P(n)$. The most useful method in vibroacoustical diagnostics, the Neyman–Pearson method (Section 4.3.3), minimizes the probability of undetected failure for an assumed or allowed level of unnecessary overhauls. Finally, the method of the maximal probability presents, on the other hand, a particular case of the method of minimum risk previously discussed, ant it is similar to the one-dimensional case of sequential analysis shown in Equation (4.23) and to Bayes's method (Equation 4.21)).

After the determination of the limit value S_l by one of the calculation techniques, we can proceed to the estimation of the condition of the object on the basis of the following simple rule distinguishing the condition on the basis of the value of the observed symptom:

$$\begin{aligned} S < S_l \quad &\text{then} \quad X = z \rightarrow \text{GOOD CONDITION} \\ S \geq S_l \quad &\text{then} \quad X = n \rightarrow \text{FAULTY CONDITION} \end{aligned} \tag{4.27}$$

So we have learned briefly the methods of making inferences about the condition of a mechanical object—methods already used, or applicable to machine vibroacoustical diagnostics. A synthetic summary of the above problem and the possibility of taking decisions is presented in Fig. 4.5, which illustrates clearly the alternatives of the diagnostic models of mechanical objects and the methods of inference behind them [49]. It follows from Fig. 4.5 that, with a high level of disturbances, the only alternatives are the probabilistic models and methods while, in low levels of disturbances, the pseudodeterministic models can be applied even with a covert dependence between condition and symptom.

In this section the concept of uncertainty or reliability of the condition estimation (good or faulty) has been discussed. This is an important problem; hence, the following section is intended to shed some light on it.

4.2.4 Accuracy of diagnostic decisions

The evaluation of the technical condition of an object carried out with the help of one of the presented methods (see Fig. 4.5) can be called the *diagnostic decision*, since it is followed be an adequate manufacturing and/or utilization procedure. However, this decision is undertaken in the conditions of uncertainty, i.e. in the presence of a random factor which may be small or presumed—as in pseudodeterministic models—or can be great and decisive—as in probabilistic models. In the latter models, the uncertainty is expressed clearly as, for example, in Equation (4.18) in the form of conditional probabilities of the GOOD CONDITION $P(z / S)$ and FAULTY CONDITION $P(n / S)$ with the observed value of symptom S.

The maximum value of the probability of the occurrence of condition $P(z / S)$ or $P(z / S)$, on the basis of which we undertake a decision, can also be called the *accuracy of condition estimation*, or accuracy of diagnostic decision. For example, if, for a given symptom, a value is smaller than the limit value $S < S_l$ and the probability of GOOD

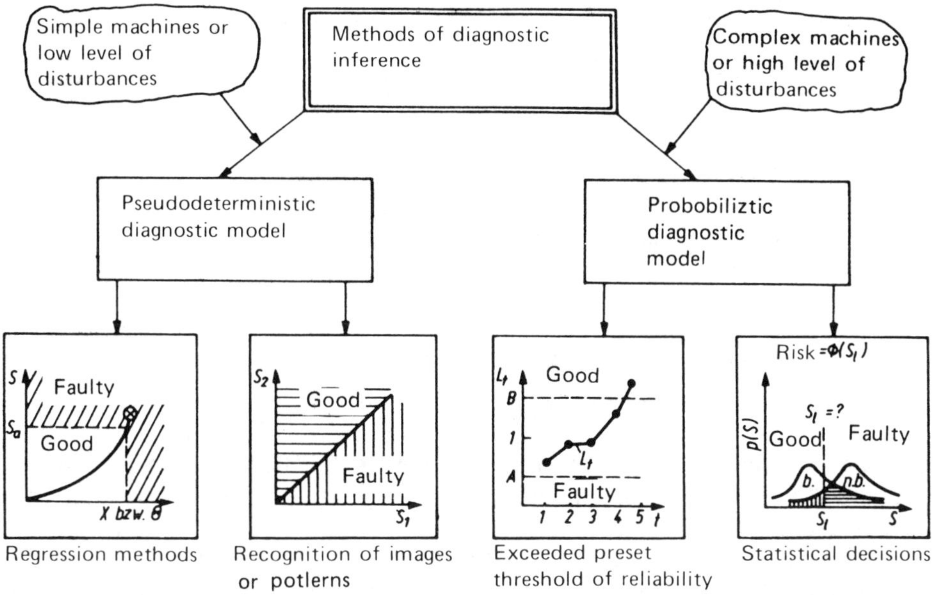

Fig. 4.5 — Synthetic presentation of diagnostic models and methods of vibroacoustical inference [49]

CONDITION is $P\ (S < S_l\ /\ z\ ) = P\ (S\ /\ \ z\ ) = 0.9$, it means that our evaluation is statistically correct in 90% of cases, and in a hundred cases of condition estimation we can make ten erroneous decisions about the object's condition. The term 'accuracy of diagnostic decisions' would be still more obvious if we knew the a priori evaluation of the good/faulty conditions $P\ (z)$ or $P\ (n)$ of the analysed group of machines and if we could compare it with the evaluation obtained after measurement which exceeds or does not exceed S_l. As we see from the above, our reasoning aims at the connection of Bayes's method of inference with the result of the theory of statistical decision in the form of the limit of symptom S_l for the evaluation of the accuracy of the decision taken on this basis.

Therefore, let us define the event of inaccurate decision about the condition, i.e. for $S < S_l$ it should be $X = z$ with the probability $P\ (S < S_l\ /\ z\ ) = P\ (S\ /\ \ z\ )$. If we carry out a measurement of symptom S on a group of machines in which the GOOD CONDITION distribution is different than the uniform one $P\ (z) \neq P\ (n) \neq 1/2$ then, according to Bayes's theorem in Equation (4.18) the final evaluation of the condition with the given symptom $S < S_l$ will be:

$$P(z\,/\,S\,) = \frac{P(z)\ P(S\ /\ z)}{P(z)\ P(S\ /\ z) + P(n)\ P(S\ /\ n)} \tag{4.28}$$

At the same time, the final evaluation of the accuracy of diagnostic decision can be taken on the basis of Equation (4.27), i.e. $S < S_l$. it follows from Equation (4.28) that the a priori information $P\ (z)$; $P\ (n)$ about the group of machines under investigation plays a role in increasing or decreasing the accuracy of the evaluation. Let us consider now what can increase our final accuracy of condition evaluation. Let us assume that the vibroacoustical symptom is a multidimensional one, and that its particular components are almost independent. Then we can make use of the evaluation of the probability of a conditional occurrence of symptom vector $\{S\}$ according to Equation (4.21):

$$P(\{S\}/z) = \prod_{j=1}^{n} P(S_j/z), \; \{S\} = \text{col}\,\{S_1,\ldots,S_n\}$$
$$P(\{S\}/n) = \prod_{j=1}^{n} P(S_j/n) \tag{4.29}$$

and after substituting Equation (4.28) we will finally obtain:

$$P(z/\{S\}) = \frac{P(z)\prod_{j=1}^{n} P(S_j/z)}{P(z)\prod_{j=1}^{n} P(S_j/z) + P(n)\prod_{j=1}^{n} P(S_j/n)} \tag{4.30}$$

Analysing Equation (4.30) for the extreme case of a uniform distribution of GOOD CONDITION $P(z) = P(n) = 1/2$, we can observe that, for the one-dimensional symptom $P(z/S) = P(S/z)$ the denominator equals 1, illustrating a certain event, but we have no advantage from the accuracy of the evaluation of the condition. On the other hand, for multidimensional symptom $n > 1$, the denominator will be smaller than 1, giving a chance of increasing the accuracy of condition evaluation.

To illustrate this possibility, let us consider the problem of vibroacoustical diagnosis of the condition of rolling element bearings with the help of the measurement of velocity $\hat{v}$ and the acceleration of vibrations $\hat{a}$. Let us consider the problem of evaluation by a single symptom $\hat{a}$ or $\hat{v}$ and the evaluation by two symptoms $\{\hat{a}, \hat{v}\}$ with different criteria for good condition of a machine group. The initial data and the calculations are illustrated in Table 4.1, quoted after another work by the Author [91].

Table 4.1 — Accuracy of the diagnostic decision for rolling element bearings at the measurement of velocity $\hat{v}$ and acceleration $\hat{a}$ [91], $I(z/\{S\})$

Good condition "a priori" assessment	Alternative measurement $\hat{v}$	$\hat{a}$	Cumulative measurement $\{S\} = \{\hat{a}, \hat{v}\}$	Possible faulty decisions for cumulative measurements
	$P(\hat{v}/z) = 0.7$	$P(\hat{a}/z) = 0.8$	$P(S/z) = 0.7 \; 0.8 = 0.56$	
$P(z) = 0.5$	0.700	0.800	0.903	~ 10%
$P(z) = 07$	0.845	0.903	0.956	~ 4%
$P(z) = 08$	0.903	0.941	0.974	~ 2%

As we see from the table, even for a uniform distribution of a GOOD CONDITION $P(z) = 1/2$, the application of both symptoms $\{\hat{a}, \hat{v}\}$ increases the accuracy of evaluation from 80% to 90%. In turn, a priori knowledge about the GOOD CONDITION of the analysed machine group can increase the accuracy of the evaluation of the condition even by 20% (from 0.7 to 0.903 for velocity $\hat{v}$). So we see that the way to increase the condition evaluation accuracy is a two-way one: (a) through a better knowledge of the analysed machine group $P(z)$; and (b) by the application of a multidimensional vector of vibroacoustical symptoms. In the latter case, an obstacle for non-critical increasing of dimensions {S} is the *cost of the diagnosis* (which has already been mentioned in the sequential method of inference). However, this is only a brief mention, and we are not going to discuss it any further here.

So far, we have analysed the total problems connected with diagnostic models and decisions referring to the evaluation of the technical condition of mechanical objects. However, there remains the question of the division into condition classes and this will be discussed in the next section.

4.3 THE LIMIT VALUE OF THE SYMPTOM AND THE CLASS OF CONDITION

4.3.1 Introduction

From the arguments contained in this book it follows that the essence of vibration diagnostics of machines consists of the comparison of the measured (on the machine) value of vibroacoustical symptom (e.g. of the vibration amplitude) to the limit value S_l and taking a utilization decision on this basis, referring to the remaining life of the machine. The limit value of symptom S_l can be determined on the basis of standards, recommendations, experiment, or even by guessing. However, a passive use of standards can lead in extreme cases to an excessively high level of *unnecessary overhauls* if the limit value of symptom is too low, or to an excessive number of *breakdowns* if it is too high. Therefore, let us look at the possible ways of creating limit value S_l , dividing the classes of the good and faulty conditions into a *two-condition classification* based on the rule of inference in Equation (4.27), correct for the regressive diagnostic model and the probabilistic model. Also, let us look at the need and possibility of creating a *multi-dimensional classification*, a three-condition classification introduced, for example, by the analogy of traffic light signalling, e.g. good condition, green; admissible, yellow; faulty, red. Such a classification of the condition facilitates the planning of overhaul works and it should be started the moment the machine enters the admissible (yellow) condition.

The considerations below leading to the determination of the limit value S_l will be divided into two cases: (a) the pseudodeterministic regressive model; and (b) the probabilistic diagnostic model.

4.3.2 Life curve of object and limit value of symptom

Let us assume that, owing to a diagnostic experiment (passive or, better, an active one) we know the life curve presenting the change of the condition of the object or the advancement of a single failure. Let the vibroacoustical image of these changes be symptom $S(\theta)$ as the function of the lifetime of object θ represented, for example, by Equation (4.6) and in Fig. 4.1 where, after the omission of the index number of symptom and the coordinate of place, we have:

$$S(\theta) = \sum_{j=1}^{n} A_j \Lambda_j(\theta) + N_\theta(\theta) \tag{4.31}$$

Here, Λ_j (θ) is the original life curve of independent failures, A_j is their weight factor of the transformation into vibroacoustical symptom, while N_θ (θ) is an uncorrelated disturbance.

Let us recall that the life curve is a deterministic or pseudodeterministic representation of condition X in the field of vibroacoustical symptom S. If we assign to condition X the lifetime θ (i.e. $X = X(\theta)$) then the life curve is a transformation of the lifetime of the object into its vibroacoustical symptom of condition. Therefore, the life curve of the object of Equation (4.31) will be obtained as the result of an active experiment carried out preferably on many objects by the measurement of the value of symptom S in the known lifetime of object θ. In conclusion, we will obtain a life curve with additional statistical information, as in Fig. 4.6.

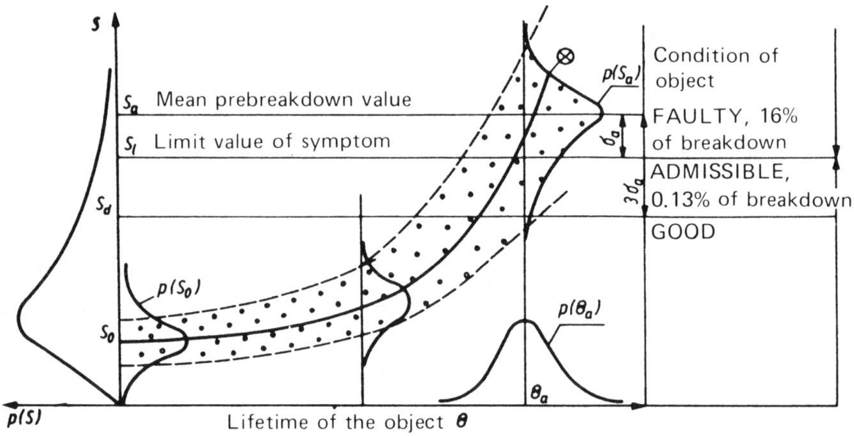

Fig. 4.6 — Life curve of object as an active experiment on many objects with the presentation of limit distributions and intermediate distributions, and the resulting proposal of condition classification

In this experiment we will obtain information about the average values of the operation time until failure θ_a, the mean prebreakdown value of symptom S_a and the mean value of the symptom of a new object S_0. The placement of the borderline of the good/faulty condition at the mean prebreakdown value would be to risk the breakdown of about half the machines going unnoticed. Therefore, it is proposed to lower the limit value of symptom S_l to:

$$S_l = S_a - \sigma_a \tag{4.32}$$

where σ_a is a standard deviation of the distribution p (S_a). If this distribution were normal (Gaussian distribution), it would still give a great probability (of the order of 16%) that the breakdown would be unnoticed. Hence, we propose to constitute the zone of the admissible condition at the limit:

$$S_d = S_a - 3\,\sigma_a = S_l - 2\,\sigma_a \tag{4.33}$$

When the machine enters such a defined admissible condition (Fig. 4.6) it signals an approaching breakdown and, according to the normal distribution of the symptom for the prebreakdown condition p (S_a), the probability of breakdown is of the order of 0.13%

while, with an unknown type of distribution from Chebyshev inequality, one can estimate the same probability at the order of 6%, which is already acceptable (see Section 4.3.3).

Summing up, one can state that in the case of an active experiment on a group of objects, one can determine from an average life curve $S(\theta)$ the limit value of symptom S_l in Equation (4.32) and, in order to decrease the number of unnoticed breakdowns, it is convenient to define the zone of the ADMISSIBLE CONDITION. Depending on the needs dictated by the utilization conditions in the plant (new or old machines, etc.) the limit value can be decreased so that $S_l = S_d$ (see Fig. 4.6).

4.3.3 Limit value in the passive diagnostic experiment

One of the techniques of inferring in the probabilistic diagnostic model (Section 4.2.3) is based on the knowledge of the limit value of symptom S_l. It is particularly helpful when a passive use of diagnostic standards (e.g. ISO-2732) does not give any good results, and it is indicated to determine independently the limit value S_l of vibroacoustical symptom based on a passive diagnostic experiment.

In order to carry out this task, let us assume that we use in a continuous operation a group of a dozen, or preferably several dozen, M machines of the same type. Among them, on the average, a number of M_z machines are in good condition (in operation) and the rest $(M - M_z)$ are under repair. The resulting estimation of availability (good condition) of the utilized group of machines is $P(z) = (M_z / M)$ and is a collective indicator of the technical condition of machines and the quality of their operation and repair.

For a group of operating machines, diagnostic measurements of vibrational magnitudes are carried out in adequate intervals of utilization time (e.g. every week or preferably in intervals of $\theta_a/20$ to $\theta_a/30$) and in adequate places, (e.g. horizontal vibration of bearing). This magnitude can, for example, be the peak vibration velocity amplitude $\hat{v}$ measured for the estimation of the general technical condition of a machine. Having at our disposal a significant number N of such observations (e.g. $N \geq 30$) of machines in different utilization conditions, we can try an independent determination of the limit value of the good condition S_l of the observed vibroacoustical symptom. The technique for the determination of this limit value based on the method of the Neyman-Pearson theory of statistical decisions is presented in other works by the present Author [89, 92], and it can be shortly explained as follows.

Let us assume that the admissible level (probability) of *unnecessary overhauls* of machines permitted by the overhaul utilization policy is A, e.g. $A = 0.05$, which means 5% of the possible overhauls. An erroneous decision about such an overhaul according to the inference rule in Equation (4.27) will take place when symptom S of a machine in good condition M_z exceeds the value S_l $(S > S_l)$. The total probability of this event is the product of the availability factor (probability of good condition) $P(z)$ and the probability of exceeding the value S_l in the GOOD CONDITION $P(S > S_l)$ (Fig. 4.7). Comparing this to the set probability of unnecessary overhauls A, we have the Neyman-pearson relation on the basis of which we can determine the optimum limit value of symptom S_l, minimizing the probability of the occurrence of breakdown:

$$P(z) \cdot P(S > S_l) = A \tag{4.34}$$

The calculation of the probability of exceeding the still unknown value S_l is not simple, but Equation (4.34) can be replaced by a simple estimation according to

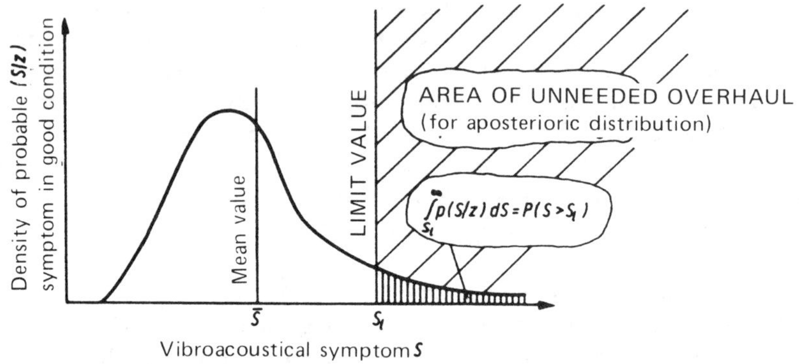

Fig. 4.7 — Illustration of the initial basis of the diagnostic probabilistic model for the determination of limit value S_l by the Neyman-Pearson technique of the theory of statistical decisions [89]

Chebyshev's inequality which finally gives [89, 92]:

$$S_l \leq \bar{S} + \sigma_s \sqrt{\frac{P(z)}{2A}} \tag{4.35}$$

where S_l is the required limit value of the symptom,

$$\bar{S} = \frac{1}{N} \sum_{i=1}^{n} S_i$$

is the mean value of the symptom in the observed group of machines M_x with N number of machine observations in different utilization conditions,

$$\sigma_s \cong \sqrt{\frac{1}{N} \sum_{i=1}^{N} (S_i - \bar{S})^2}$$

is the assessment of standard deviation of observations,

$$P(z) = M_z / M$$

is the evaluation of the probability of good condition (availability) of the observed machine group, and A is the probability of unneeded overhauls (erroneous diagnostic decisions) admissible in the given plant, $0 < A \ll 1$. Therefore, the limit value S_l follows directly from the properties of the observed group of machines and depends on their general condition, the quality of overhauls ($\bar{S}$, σ_s, $P(z)$) and the assumed level of unnecessary overhauls $A \ll 1$.

For a group of new machines put into operation with a high quality of overhauls, the mean value of the vibration symptom $\bar{S}$ and its dispersion σ_s will be small, while the availability $P(z)$ will be high. Therefore, we will require a low level of overhauls which, in effect, will give an increased limit value of symptom S_l in Equation (4.35). A similar transfer from the system of planned preventive overhauls to the system of overhauls on the basis of the knowledge of the real technical condition evaluated with the help of the measurement of symptom S will give, in effect, a gradual increase of $P(z)$ and σ_s with a slight increase $\bar{S}$. So, with the help of Equation (4.36) we can gradually *set the limit*

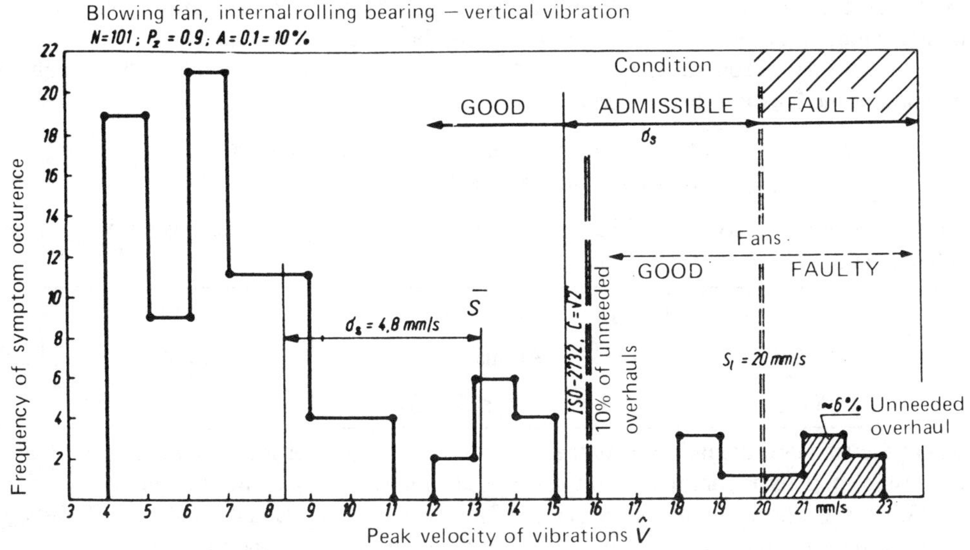

Fig. 4.8 — Histogram of the distribution of the peak amplitudes of vibration velocity of blowing fans WDR 1700/1800 [94]

value of S_l during the implementation of vibroacoustical diagnostics in the industrial plant.

For a better understanding and presentation of the above techniques for the determination of S_l let us consider an example taken from a real case of diagnosing blower fans of a power unit [93, 94] which is presented in Fig. 4.8. It can be seen in the figure and in many others [93, 94] that the histogram of the distribution of amplitudes is skew-asymmetric which, after all, results from the laws of transferring the life curve into the field of symptom in a passive diagnostic experiment (see the total distribution density for machines in operation, e.g. in Fig. 4.6, being there the density of the GOOD CONDITION $p\,(S/z)$).

To the calculation of the limit value with the data $N = 101$ observations carried out every 2 weeks on twelve objects with an availability $P\,(z) = 0.9 = 90\%$, we have assumed an admissible level of unnecessary overhauls $A = 0.1 = 10\%$, since the fans had a very long total working time and short average time between failures of the order of several months. It follows from the observation that the only significant reason for failure was the unbalance of the rotor which gives a sinusoidal character of the time domain of vibration velocity. Therefore, we can assume that the peak coefficient $\sigma = (\hat{v} / v_{\mathrm{rms}}) = 2$ and we can convert the standard limit value ISO-2732 group G with $v_{\mathrm{rms}\,l} = 11.8$ mm/s to the peak limit value $\hat{v}_l = C\, v_{\mathrm{rms}\,l} = 15.8$ mm/s (see also Fig. 5.4(a)).

On the basis of the estimation of the limit value (Equation (4.35)) we have S_l = 20mm/s, from the data in Fig. 4.8, which gives the real level of unnecessary overhauls at ~6%. On the other hand, according to ISO recommendations, we obtain 10% of unnecessary overhauls (Fig. 4.8). Therefore, we can state that we have obtained an important technique for estimating the limit value S_l, taking into consideration the real operational conditions in an industrial plant. Therefore, in the determination of S_l, we can consider the objective properties of the utilized machine group known from the passive experiment (P (z), $\bar{S}$, σ_s) as well as subjective premises of the utilization policy of the management of the plant contained in the admissible level of overhauls A.

As the final result of the presented experiment in Fig 4.8, we see the resulting division of the fans into two classes:

$$S < S_l = 20 \text{ mm/s} \rightarrow \text{fans in the GOOD CONDITION}$$
$$S \geq S_l = 20 \text{ mm/s} \rightarrow \text{fans in the FAULTY CONDITION}$$

Also, a three-class division of fan condition has been proposed here according to the scheme below:

	$S_l - \sigma_s$		S_l		
GOOD	|	ADMISSIBLE	||	FAULTY	(4.36)
$S < S_l - \sigma_s$	|	$S_l - \sigma_s \leq S < S_l$	||	$S \geq S_l$	

where the standard deviation σ_s is the measure of the width of the interval of uncertainty (admissible). This recommendation can appear contradictory to the recommendation in Section 4.3.2, where the proposed width of interval was $3\sigma_a$ (Equation (4.33)). However, as we remember, the point of departure was the distribution of the prebreakdown values of symptom p (S_a). On the other hand, we base the whole distribution on the machine population in good condition p (S/z) with significantly higher standard deviation since $\sigma_s > \sigma_a$ (see left-hand side of Fig. 4.6).

We can recommend the above passive experiment with Equation (4.35) which, with a cost of labour and calculation takes into consideration the majority of specific features essential for the given industrial plant. The approach proposed here is not the only one, and in this Author's research team the discussion is continued and alternative methods are considered based on a passive experiment and a partial knowledge of the life curve. It is a little more complicated and so far, has not been finalized; hence, we refer the interested reader to an example of the works by Nowicki and Sordyl on this subject [95]. As was shown recently, much better results can be obtained with Pareto symptom distribution which enables us also forecasting of residual time to failure.*

4.4 SIMPLE AND GENERALIZED CONDITION SYMPTOMS—FILTRATION OF DISTURBANCES

4.4.1 Introduction

The final stage of diagnostics is the decision-taking process about the condition of the object on the basis of a comparison with the standard measured values of vibroacoustical symptoms of machine condition. However, these symptoms, next to the useful information about the condition of the object, carry a whole range of information unnecessary from the point of view of diagnosis, i.e. disturbances. Therefore, we have to foresee some minimization measures of the influence of these disturbances on the reliability of diagnostic decision-making consisting of the differentiation of conditions (e.g. good/faulty) of the mechanical object. From the utilitarian point of view, we can distinguish the two most important tasks of the discrimination of the objects condition as follows.

* Cempel (1991) Condition evolution of machinery and its assessment from a passive diagnostic experiment. *Mechanical Systems and Signal Processing*, **5** (4), pp.317–326.

(1) We have a group of objects with the same lifetime and it is necessary to differentiate their condition (for θ = const), or quality (for $\theta = 0 \Rightarrow$ new objects). At the same time, measurement data are available in the form of a set of symptom values $S(r, \theta)$ measured in different points 'r' of the object.
(2) We observe one object during operation when θ = var, while the task consists of finding the change in its condition. Similarly, we have at our disposal a set of vibroacoustical symptoms measured at different points of the object.

The difficulty in taking the proper diagnostic decision in each case follows from the existence of random disturbances contained in each fact of measurement (observation) of vibroacoustical symptom. It occurs even in the simplest case of one symptom $S(r, \theta)$, and even for one failure with a life curve $F(\theta)$. In more complex cases of many independent failures with life curves $F_i(\theta)$, $i = 1, \ldots, n$, the measurement disturbances following from the unknown control and unknown interaction with the surround are additionally accompanied by our ignorance of the way of the object's wear. Also, we do not know which failure $F_i(\theta)$ is the dominating one and, therefore, we do not know which failure-oriented discriminating factor should be measured in order to get a correct evaluation of the condition. We will try to alleviate this lack of knowledge by performing the procedure of filtration and smoothing these disturbances. These procedures are not obvious [96]; therefore, we will discuss them in the next Section 4.4.2, illustrating them with examples taken from diagnostic practice.

4.4.2 Linear averaging in respect of utilization time and the object's life

The presentation and motivation of an effective method of disturbance filtration requires a clear diagnostic model of the observation of a mechanical object, a model that would indicate the transformations of the advancement of the particular failures in the machine $F_i(\theta)$ into a vibroacoustical signal, and then into vibroacoustical symptom owing to an adequate system of observation. In looking for a sufficiently general diagnostic model, let us return to the considerations from Section 3.2, which presented the diagnostic model of the observation of a complex mechanical object with n-independent failures described by life curves $F_i(\theta)$, where $i = 1, \ldots, n$. These functions $F_i(\theta)$ in the exemplary object reflect, for example the advancement of: (a) the wear process of the rolling element bearing; (b) cavitational erosion of the rotor; (c) fatigue failure of the blade; and (d) other independent failures. Our object is also sufficiently extended so that by measuring the same vibroacoustical symptoms in different places of the object (the coordinate r) we record the same symptoms of wear $F_i(\theta)$ but with different relations of the useful signal to the disturbing noise. Such a model is presented in Fig. 4.9 taken from a previous work [96]; at the same time it corresponds with considerations from Section 3.2. As we can see from the figure and Equation (3.2), the generated vibroacoustical signal is interchangeable with the advancement of failures $F_i(\theta)$ through the original signals of failure $|V_i \cdot (t, \theta)| \sim F_i(\theta)$ and is disturbed by noise $n(r, t, \theta)$. Finally, the vibroacoustical signal can be written in the form:

$$s(r, t, \theta) = \sum_{i=1}^{n} h(r_i, t, \theta) * V_i(t, \theta) + n(r, t, \theta) \tag{4.37}$$

where $*$ denotes the operation of the convolution of the impulse characteristics of the system $h(r_i, t, \theta)$ with the original signal $V_i(t, \theta)$.

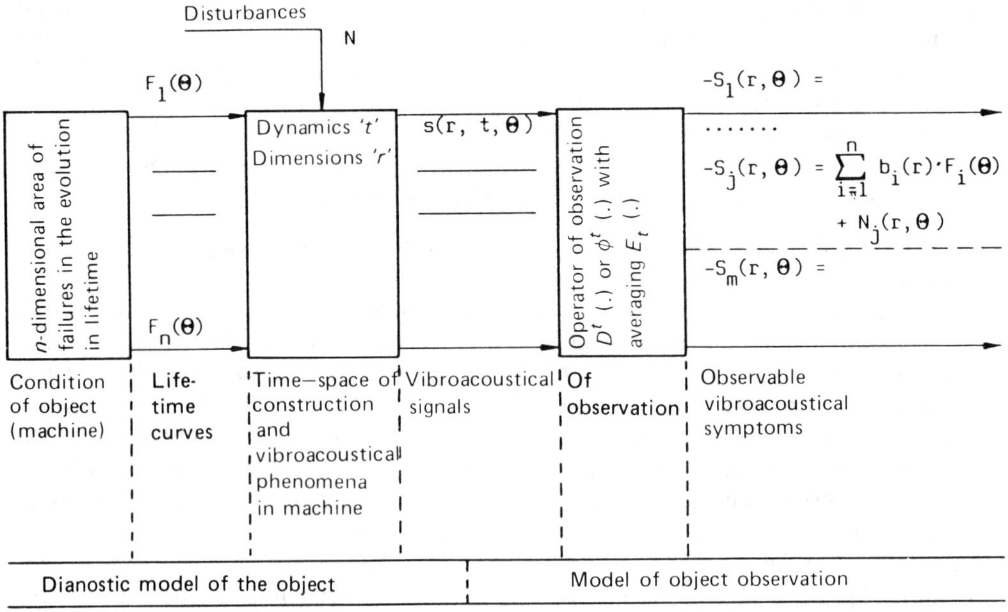

Fig. 4.9 — Model of diagnostic observation of a mechanical object with n-dimensional space of failures [96]

The system of observation visible in Fig. 4.9 permits us to transform the signal in Equation (4.37) into the measurement values interchangeable with the condition of the object (i.e. vibroacoustical symptom with the help of operation $\phi^t(\cdot)$ and averaging $E_t(\cdot)$ in respect to dynamic time t) while discriminants with the help of operation $D_k^t(\cdot)$ and averaging.

These operators are presented in Equations (3.3)–(3.7). The result of the observation of operator's action can be written as follows.

Vibroacoustical symptoms:

$$S_j(r,\theta) = \sum_{i=1}^{n} b_i(r)\, F_i(\theta) + N_j(r,\theta), \quad j = 1,\dots, m$$

Discriminants (4.38)

$$D_i(r,\theta) = b_i(r)\, F_i(\theta) + N_i(r,\theta), \quad i = 1,\dots, n$$

It can be seen that, in both cases, there occurs a noise component $N_j(r, \theta)$ whose mean-square amplitude can be of the order of the determined part:

$$\left[N_j(r,\theta)\right]_{\mathrm{rms}} = \left[\frac{1}{\Delta}\int_{\theta=\Delta/2}^{\theta+\Delta/2} N_j^2(r,\theta_0)\, d\theta_0\right]^{1/2} \le \sum_{i=1}^{n} b_i(r)\, F_i(\theta)$$

or in the categories of signal to noise relation:

$$\frac{S}{N} = \frac{\sum_{i=1}^{n} b_i(r)\, F_i(\theta)}{\left[N_j(r,\theta)\right]_{\mathrm{rms}}} \geq 1 \tag{4.39}$$

where Δ is the interval of averaging time smaller than the limit time of breakdown $\Delta < \theta_{aw}$.

In addition, assume that they are random disturbances in the fields r and θ with a zero mean value, i.e.:

$$E_\theta\{N_j(r,\theta)\} = 0, \qquad E_r\{N_j(r,\theta)\} = 0, \; j = 1, \ldots, m$$

and that in both fields they are mutually uncorrelated:

$$E_\theta\{N_u(r,\theta)\, N_v(r,\theta)\} = 0, \qquad E_r\{N_u(r,\theta)\, N_v(r,\theta)\} = 0, \; u \neq v \tag{4.40}$$

where:

$$E_\theta\{\cdot\} = \frac{1}{\theta_{aw}} \int_0^{\theta_{aw}} (\cdot)\, d\theta, \qquad E_r\{\cdot\} = \frac{1}{\upsilon} \int_v (\cdot)\, d\upsilon$$

are operators of the mean value with respect to the utilization time θ and the volume υ of the object.

Proceeding now to the search for a method of increasing the signal to noise ratio, let us first notice the current average also called the moving average which is applied frequently in econometrics [97] and, in a slightly different form, is used in the analysis of time series [98]. Averaging, according to the assumptions about zero mean value, can be applied to the field of time θ, smoothing the life curve and thereby minimizing the disturbances.

Determining the current averaging operator by sign b, it can be expressed in an integral form for symptom $S_j(r,\theta)$ in the following way:

$$E_\theta^b\left[S_j(r,\theta)\right] \equiv \frac{1}{\Delta} \int_{\theta-\Delta/2}^{\theta+\Delta/2} S_j(r,\theta_0)\, d\theta_0 = S_j^*(r,\theta)$$

or in a variable discrete $w + 1$-point form:

$$E_\theta^b\left[S_j(r,\theta_u)\right] \equiv \frac{1}{w+1} \int_{t=u-w/2}^{t=u+w/2} S_j(r,\theta_t)\, d\theta_0 = S_j^*(r,\theta_u) \tag{4.41}$$

where θ_t denotes the successive equally spaced time points of the measurement of vibroacoustical symptom $t = 0, 1, \ldots$, while $w + 1$ is the odd number of points taken for averaging, usually $w = 3$ or, less frequently, $w = 5$.

In a definite case of a three-point averaging, the successive values of the smoothed realization of life curve can be found from:

$$S_j^*(r,\theta_u) = \frac{1}{3}\left[S_j(r,\theta_{u-1}) + S_j(r,\theta_u) + S_j(r,\theta_{u+1})\right] \tag{4.42}$$

$$u = 0, 1, \ldots, P$$

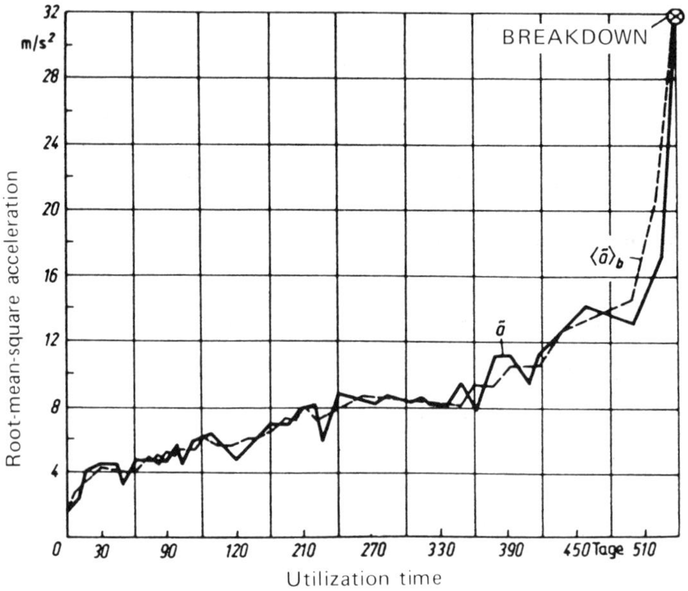

Fig. 4.10 — Three-point current averaging $\langle\bar{a}\rangle_b$ in comparison with the course of root-mean-square vibration acceleration of an electric motor SE1 [96]

and at the same time in the calculation of the initial point $u = 0$, the first part of Equation (4.42) is zeroed and the last term is zeroed in the calculation of the final point.

The minimum disturbance properties of the averaging operation in Equations (4.41) and (4.42) have been given [96], and we are going now to consider an example of the action of the current averaging. The effect of the minimization of disturbances in the life curve of a rolling element bearing for an electric small power motor by way of current averaging is presented in Fig. 4.10. We see here the significant smoothing of the realization of the life curve and we infer that for such symptoms as a root-mean-square vibration, acceleration amplitude, and for multipoint and dense averaging it will have an exponential character.

The application of averaging in the field of lifetime θ for the minimization of disturbances, is the simplest means, though not the only one. From diagnostic practice, and from the model in Fig. 4.9 and Equation (4.38), there follows an equal change of condition F_i (θ) for all points in the object's area with random participation of disturbances $N_j(r, \theta)$. Therefore, we will apply the operation of averaging with respect to the area of the object for the minimization of disturbances. In the continuous-integral form this operation of the symptom in Equation (4.38) will have the form:

$$
\begin{aligned}
E_r\{S_j(r,\theta)\} &= \left\langle S_j(\theta)\right\rangle_r \equiv \frac{1}{\upsilon}\int_\upsilon S_j(r,\theta)\,d\upsilon \\
&= \sum_{i=1}^{n}\frac{F_i(\theta)}{\upsilon}\int_\upsilon b_i(r)\,d\upsilon + \frac{1}{\upsilon}\int N_j(r,\theta)\,d\upsilon \qquad (4.43)\\
&= \sum_{i=1}^{n}\bar{b}_i F_i(\theta) + \overline{N_j(\theta)}
\end{aligned}
$$

The discrete version of such averaging with respect to the area of an object with M points of measurement will be more readable, since:

$$\begin{aligned}\left\langle S_j(\theta)\right\rangle_r &= \frac{1}{M}\sum_{u=1}^{M} S_j(r_u,\theta)\\ &= \frac{1}{M}\sum_{u=1}^{M}\left[\sum_{i=1}^{n} b_i(r_u)\,F_i(\theta)\right]+\frac{1}{M}\sum_{u=1}^{M} N_j(r_u,\theta)\\ &= \overline{N_j(\theta)}+\sum_{i=1}^{n}\overline{b_i}\,F_i(\theta)\end{aligned} \tag{4.44}$$

As we see from the above, both the continuous and the discrete versions refer to the same utilization time θ. Therefore, this type of averaging can find applications in vibro-acoustical-acoustical investigations of the quality of new objects $\theta = 0$, or in trials to distinguish the technical condition of objects in the same utilization time $\theta \neq 0 = \text{const}$. The application of the volumetric averaging in respect to eighteen measurement points of the stand for the examination of bearings for one symptom, and the product and quotient of symptoms is shown in Fig. 4.12.

As we see from this example, the best discriminating property of the condition of head bearings, among the techniques of the linear averaging, has the volumetric averaging of the products of symptoms: $\frac{1}{18}\sum_{1}^{18}\hat{v}_i\,\tilde{v}_i$ and is competitive with geometrical averaging: $\prod_{1}^{18}\hat{v}_i$ which will be discussed in Section 4.4.3. So we see that such a simple operation of averaging in respect to volume has significant properties of disturbance filtration of the measurement symptoms, and it is worth remembering this in the problem of technical condition comparison of similar objects for $\theta = 0$ and $\theta = \text{const} \neq 0$.

4.4.3 Geometrical averaging—cumulants

Averaging of one symptom with respect to the span of the object can be carried out also in another way, namely by geometrical averaging used in econometrics [86, p. 326]. In mathematical categories, the geometrical or harmonic average of order M is the M-th radical from the product of these values. In vibroacoustical diagnostics, a magnitude calculated in this way can serve two purposes: (a) distinguishing the vibroactivity of a mechanical object; and (b) for the discrimination of the dominating mechanism of getting damaged $F_i(\theta)$ [96].

In order to illustrate the above objective of the vibroactivity of objects, let us imagine that symptom $S_j(r_u, \theta)$ measured in object points r_u, $u = 1, \ldots, M$ is, for example, a root-mean-square or peak vibration velocity. The product of these velocities will be analogous to the volume stretched on these vectors and, at the same time, be a measure of the vibration activity of the object, i.e. a measure of energy dissipated by vibrations. Denoting this value by $W(\theta)$ we have:

$$W(\theta) = \prod_{u=1}^{M} S_j(r_u,\theta) = \prod_{u=1}^{M}\left\{\sum_{i=1}^{n}\mathbf{b}_i(r_u)\,\mathbf{F}_i(\theta)+\mathbf{N}_j(r_u,\theta)\right\} \tag{4.45}$$

we assume at the same time that the particular components are non-zero ones. On the other hand, when we take the M-th root from the energy dissipated by vibrations, we have in turn the geometrical mean of our symptom $S_j(r, \theta)$ called also the *cumulant* since:

$$\Pi_M(\theta) = \sqrt[M]{W(\theta)} = \left[\prod_{u=1}^{M} S_j(r_u, \theta)\right]^{\frac{1}{M}} \tag{4.46}$$

The latter name, cumulant, becomes more justified when the product $W(\theta)$ is created from different symptoms or discriminants [96] reflecting the different mechanisms of wear. Then the value $\Pi_M(\theta)$ cumulates all their properties and reflects globally the wear process of the object.

We can also show that the geometrical averaging lowers the level of disturbances $N_j(r, \theta)$ in relation to the determined part $\Sigma\, b_i(r)\, F_i(\theta)$ informing about the condition of the object. Therefore, we can say that the vibroactivity will be the distinguishing of the technical condition of the examined object for θ = const.

To illustrate these properties, let us take into consideration [96] the vibrations of the gear wheel bearings of a brown-coal excavator. Such an object is a large one (about 6 × 3 × 2 m) and it is necessary to compare the technical condition of three gears and monitor the working condition of one of them.

An attempt at a comparison of the condition of the gears by the measurement of the peak acceleration of the bearings of rollers A–E is presented in Fig. 4.11 (a). As we see from the figure, the comparison of the accelerations of the particular bearing casings does not give any unmistakable indications regarding the general technical condition of the particular gears. On the other hand, the calculation of the cumulant $\Pi_M(\theta) = \sqrt[M]{W(\theta)}$ (logarithmic scale was applied here) showed the gear SRS 1200 III to be the worst one. One of these gears (SRS 1200 II) was vibrationally observed for several months in order to develop a vibrational method of condition estimation which is presented in Fig. 4.11(b).

As we see from these investigations, the object is really extensive, since the breakdown condition of bearing E is not observed in bearings A, B or C. Furthermore, the replacement of the bearing did not improve the general technical condition of the gear reflected by the cumulant $W(\theta)$. However, the replacement of the bearing decreased the degradation rate.

The above experiment suggests two conclusions: (a) the vibration level should be observed in all essential construction nodes; and (b) the cumulant of the observed vibration values can be used as the measure of the general technical condition of the object giving the information about the general level of its degradation connected with one failure or with many different failures. For an analytical signalling of such properties, let us assume that in many points of object r_u, $u = 1, \ldots, M$, we have measured the value of the symptom oriented to the j-th failure (i.e. the discriminant $D_j(r_u, \theta)$ which can be approximated by the exponent with intensity λ_j. Then instead of Equation (4.38), we have:

$$D_j(r_u, \theta) = N_j(r_u, \theta) + B_j e^{\lambda j \theta} \tag{4.47}$$

Calculating the cumulant of the above in relation to M measurement points we have:

$$\Pi_M(\theta) = \left[\prod_{u=1}^{M} D_j(r_u, \theta)\right]^{1/M} = B_j e^{\lambda j \theta}\left[\prod_{u=1}^{M}\left(1 + \frac{N_j(r_u, \theta)\, e^{\lambda j \theta}}{B_j}\right)\right]^{1/M} \tag{4.48}$$

$$\cong B_j e^{\lambda j \theta} \quad \text{for } \theta > \theta_0$$

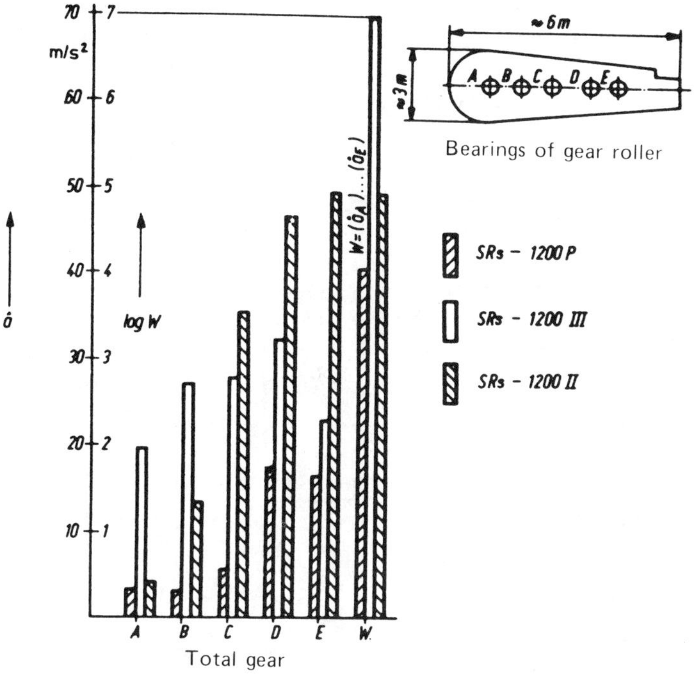

Fig. 4.11(a) — Peak values of vibration accelerations $\hat{a}$ of the gear wheel of excavator SRS-1200 II compared with other gears in the particular measurement points and the image of the cumulant W [96]

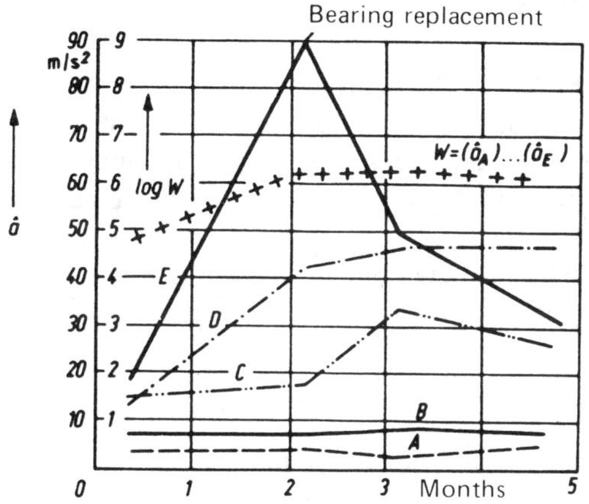

Fig. 4.11(b) — Change of peak value $\hat{a}$ of vibration acceleration over time for the same gear as above.

Therefore, we see that by this method of object observation the dominating *j*-th failure becomes dominant in our cumulant for sufficiently long utilization times θ_0.

As an example of such behaviour of the cumulants, let us take into consideration Fig. 4.12, which presents attempts to distinguish vibroacoustically the qualities of four

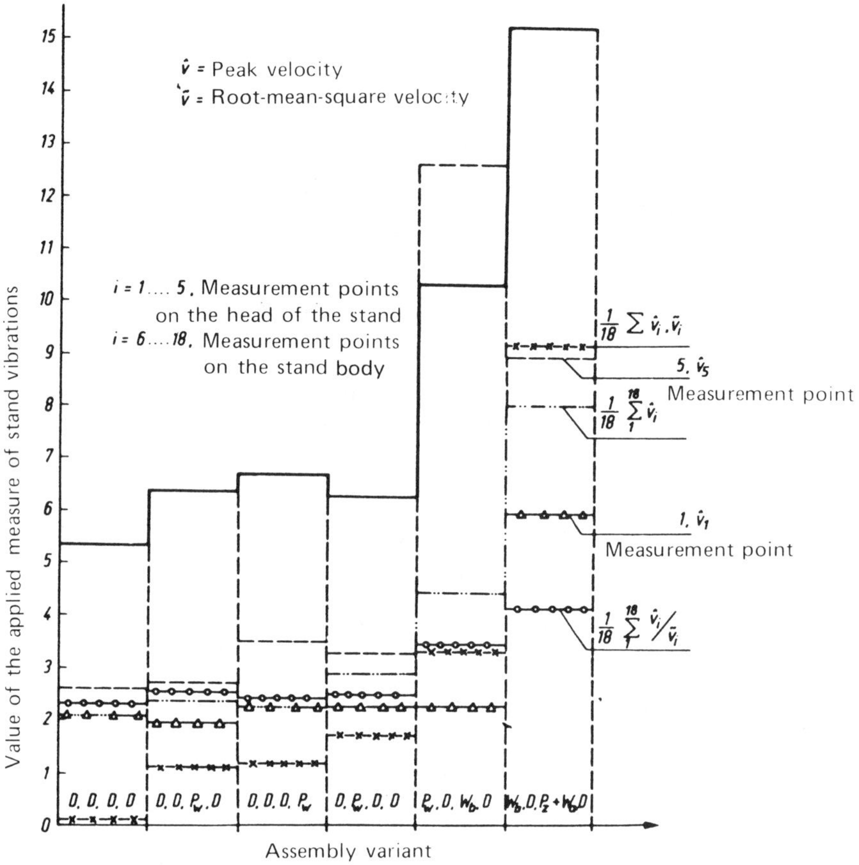

Fig. 4.12 — Possibilities of distinguishing the condition with the help of simple and cumulative measures of the test stand vibrations. Combination of four bearings in the head: D signifies good, P_w means the internal ring is impaired, P_z means the external ring is impaired and W_a, W_b means roller a or b is impaired [96]

bearings mounted in the head of the stand for a durability test. Hence, we see that the best dynamics and condition discriminating ability is in the measure

$$\log\left(\prod_{i=1}^{18} \hat{v}_i\right),$$

where v_i is the peak vibration velocity of the i-th point, being a modified cumulant (Equation (4.48)) where the logarithm was used instead of the radical. Other methods of the arithmetical averaging in respect to the span of the object applied here do not possess this discriminating power, although in comparison with the course of velocity in the measurement point 1, $\hat{v}_1$, the volumetric averaging gives better results.

The averaging of one symptom in respect of the span of the object or in respect of the time of utilization does not always give satisfactory results in the recognition of the condition or the quality of the object. There are two reasons for such a situation. The first is the already mentioned spatial extension of the object, where with the change of the place of measurement there comes a significant change of the relation between the signal and the noise (see, for example, Fig. 4.11(b), where the failure of bearing E is not visible in the majority of measurement points). On the other hand, the majority of the presently used vibroacoustical symptoms have a similar character to the discriminants preferring one fault before another one. For example, the measurement of accelerations is applied in the diagnostics of rolling element bearings and transmission gears, while for the evaluation of the condition of journal bearings the vibration displacements are used. Hence, for the detection of most failures in the object, a multipoint measurement of several different vibroacoustical symptoms is necessary.

Being aware of such a need, let us extend the definition of the cumulant in Equations (4.45) and (4.46) to different symptoms $S_j\,(r_u,\theta)$, $j = 1, \ldots, m$, measured in different places of the object $u = 1, \ldots, M$.

$$W_m(r_u,\theta) = \prod_{i=1}^{m} S_j(r_u,\theta) = \prod_{j=1}^{m}\left[N_j(r_u,\theta) + \sum_{i=1}^{n} b_i(r_u)\,F_i(\theta)\right] \tag{4.49}$$

Such a measure has been defined by the Author [7, Section 8.6.2] where it was additionally subject to current averaging $E^b_\theta\,\{W_m\,(r_u,\,\theta)\}$ and it was called the *moment of the* m-*th order.* As an example of such a procedure in relation to the originally observed vibroacoustical symptoms of the condition of an electric motor $\tilde{a}$, H_v, f_v is presented in Fig. 4.13 [96]. Here, $\tilde{a}$ is the root-mean-square vibration acceleration of the electric motor bearing, f_v is Rice's frequency of these vibrations, H_v is the harmonicity factor of these vibrations, and $W_3(\theta)$ represents their product smoothed in three points.

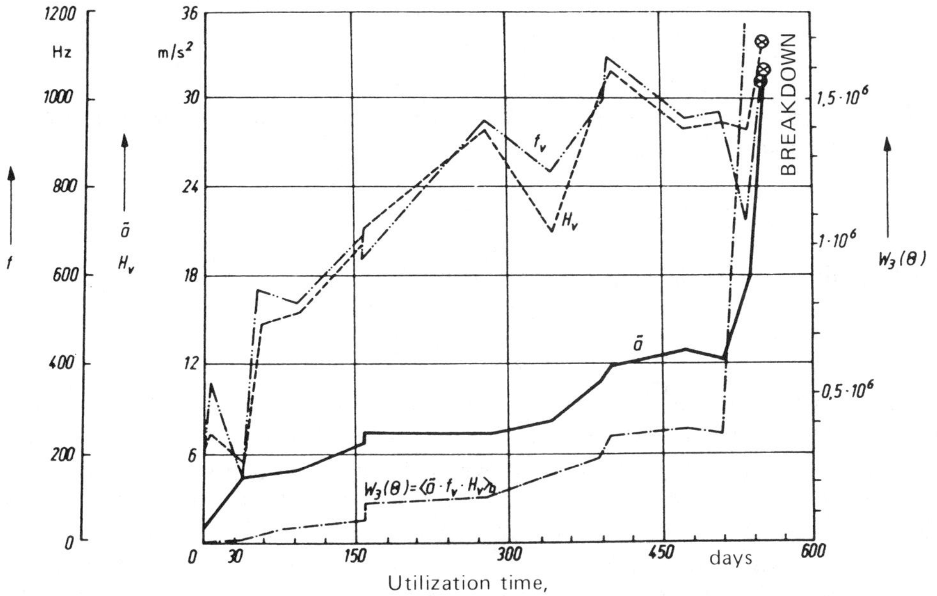

Fig. 4.13 — Trend of root-mean-square acceleration $\tilde{a}$, Rice's frequency of velocity f_v, coefficient of harmonicity H_v, and their current moment averaged in three points $W_3(\theta)$ for an electric motor SE2 (bearing vibrations) [96]

Inspecting Fig. 4.13, one can say that the product cumulative measure presents, in a better and more stable way, the change in the technical condition of the bearing.

A similar diagnostic application is presented in Fig. 4.14 according to he data of vibration velocity in three directions (vertical, horizontal and axial) of a journal bearing of a turbine-driven pump [96]. At the same time, the use of linear and logarithmic scales has been compared for a better representation of the trend of changes in the condition of the unit. The logarithmic scale shows small interferences of the life curve; on the other hand, the averaged life curve $W_3(\theta)$ (cumulative product measure) valued in the linear scale gives the most readable picture of the change in the technical condition of the turbine-driven pump.

Concluding the considerations on the minimization of the disturbances of symptoms, we will show that the product operation itself, or the cumulative measure which is part of the geometrical mean, has significant disturbance minimizing properties. It is most easily demonstrated in the example of the products of discriminants $D_j(r_u,\theta)$ oriented towards failure. Creating a cumulant from the exponentially variable discriminants in Equation (4.47) we will have:

$$W_m(r_u,\theta) = \prod_{i=1}^{m} D_j(r_u,\theta) = \prod_{j=1}^{m}\left[N_j(r_u,\theta) + b_j(r_u)\, e^{\lambda j\theta}\right]$$

$$= \exp\left(\sum_{j=1}^{m} \lambda_j \theta\right) \prod_{j=1}^{m} b_j(r_u)\left[1 + \frac{N_j(r_u,\theta)\, e^{-\lambda j\theta}}{b_j(r_u)}\right] \tag{4.50}$$

$$\cong \exp\left(\sum_{j=1}^{m} \lambda_j \cdot \theta\right) \prod_{j=1}^{m} b_j(r_u), \quad \text{for } \theta > \theta_0$$

As we see from the above, for sufficiently long times, $\theta > \theta_0$, the operation of discriminant product minimizes the disturbances $N_j(r_u, \theta)$, leaving only the total life curve of all failures of the machine,

$$\exp\left(\sum_{j=1}^{m} \lambda j \theta\right)$$

scaled by the corresponding coefficient

$$\prod_{j=1}^{m} b_j(r_u).$$

It is easy to guess that further minimization of disturbances can be obtained by current averaging of the cumulant in Equation (4.50) by the operation $E^b_\theta\{\cdot\}$ and by the linear or geometrical averaging with respect to the span of the object (Fig. 4.12).

Summing up the above considerations we can say that in the minimization of symptom disturbances we can use the current averaging with respect to the utilization time θ, the simple arithmetical averaging in respect to the span of the object r, and the geometrical averaging with respect to the same domain. The latter case used with different symptoms or discriminants gives a new cumulative measure of the moment type which can carry combined information about the advancement of all failures of the machine. This type of additional properties of the cumulants is in many cases extremely important, since it reduces the multidimensional area of failures to one dimension (see Equation (4.50)). It gives further extreme simplification of the methods and technique of inference where, for

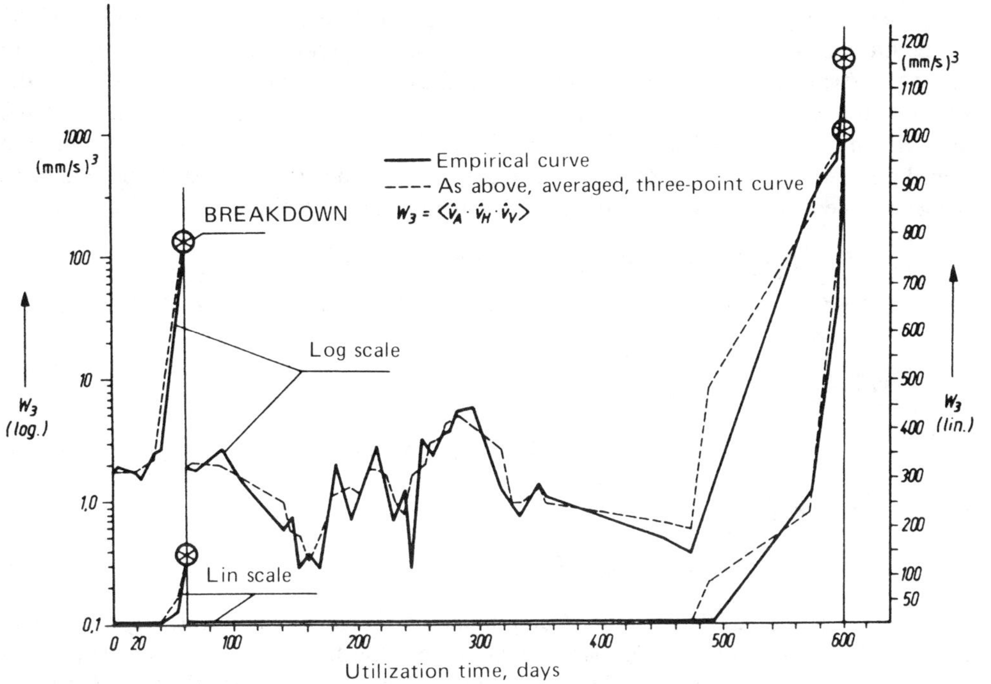

Fig. 4.14 — Original and averaged life curve of a turbine-driven pump according to a generalized measure W_3 [96]

example, instead of the complicated technique of pattern recognition we can use the Neyman-Pearson technique described in Section 4.3.3. It gives additional smoothing of the cumulant in comparison with the course of the initial symptoms in the scale, since the operation of creating them possesses disturbance-minimizing properties.

4.5 CONDITION PROGNOSIS

4.5.1 Introduction

In our outline of the knowledge and ability of vibroacoustics diagnostics we know: (a) which signal to measure; (b) how to transform it into a vibroacoustical symptom; (c) how to determine the usability limits regarding the vibroacoustical symptom; and (d) how to minimize the influence of incidental interferences. Therefore, we are able to determine the technical condition of the object, i.e. we are able to make a diagnosis. We also can estimate the causes and origin of the definite condition with the help of the spectral, cepstral and product spectrum analyses, etc. The only unsolved problem being part of the diagnosis in its full understanding is the *prognosis* of the future condition. This section will be devoted to the clarification of this problem.

The paradigm of diagnostics, as well as vibroacoustical diagnostics, is the *causality* and, sometimes, even the determinism of the relation condition → symptom, or *wear advancement → symptom.* If so, it should be possible to carry out the prognosis of the condition with the help of the available models of the trend condition → symptom. In order to explain this problem in vibroacoustical diagnostics, we will clarify it from the very beginning (i.e. from the moment of the reception of vibroacoustical signal $s\ (r, t, \theta)$) in a definite place r of the object, where we have decided to mount the vibration transducer. In this place, the character of the signal and the condition–symptom relations are determined by the following differential equations and definitions:

vibroacoustical process interchangeable with signal $s\ (r, t, \theta)$:

$$m(X, \theta)\, \ddot{s}(r, t, \theta) + c(X, \theta)\, \dot{s}(r, t, \theta) + k(X, \theta)\, s(r, t, \theta) = f(X, \theta, Z, g, t) \tag{4.51}$$

condition $X\ (\theta)$:

$$\frac{d X(\theta)}{d \theta} = \phi(s, \theta, g) \tag{4.52}$$

observed symptom $S\ (r, \theta)$:

$$S(r, \theta) = E_t\{\psi[s(r, t, \theta)]\} = L(r, X) + N(r, \theta) \tag{4.53}$$

where $m\ (\cdot)$, $c\ (\cdot)$ and $k\ (\cdot)$ are mass, damping and stiffness of the object, respectively, in place of the reception of the vibroacoustical process transformed into signal $s\ (\cdot)$; $f\ (\cdot)$ is the forcing of the object depending on its condition $X\ (\theta)$, its disturbances Z, and its loading by the technological process g; $E_t\ \{\cdot\}$ is the known operator of averaging in respect to the field of dynamic time; $\psi\ [\cdot]$ is the operator creating the vibroacoustical symptom; $L\ (r, X)$ is the deterministic part of symptom (trend); and $N\ (r, \theta)$ denotes the disturbances randomly present in the observed vibroacoustical symptom.

Looking at mutual dependence in the relations (4.51)–(4.53), the simplest formal method of solution should consist of the operational solution of the condition Equation (4.52] and the introduction of a signal into Equation (4.51). The formal solution into the equation permits the introduction of the solution into the equation of observation (Equation (4.53)). Hence, we see that the that the formal solution (Equation (4.52)) makes $L\ (\cdot)$ an integrodifferential operator between the symptom and the condition. Whether this is the case or whether $L\ (\cdot)$ can be a function depends on the dominating mechanism of wear in the object and on our knowledge about the phenomenon. For example, the condition–symptom relations for fretting and the frictional wear process will have a functional character by analogy to Equations (2.21) and (2.22); on the other hand, for fatigue wear the development of the gap is described by the differential Equation (2.16). In this connection the condition–symptom relation will be simpler and defined in the operational field.

Thus $L\ [r, X\ (\theta)]$ can be, in general, the integrodifferential operator or even a differential operator acting on condition $X\ (\theta)$ in relation to the time of utilization θ. In simplified cases, it will be only a function of place r and condition $X\ (\theta)$.

Thus, we have a causal connection between the condition and the observed symptom in the form:

$$S(r, \theta) = L[r, X(\theta)] + N(r, \theta) \tag{4.54}$$

which, depending on the amplitude relation between $L(\cdot)$ and the disturbances $N(\cdot)$ can be pseudodeterministic for small $N(\cdot)$, and probabilistic for greater level of disturbances. In the first case of small disturbances the connection condition–symptom can be easily found, and therefore we can define value $S(\cdot)$ for optional θ and optional $\Delta\theta$ in advance from the moment of the last observation. In the second case of significant disturbances, $N(\cdot)$, we only know that there is a causal connection, but its operational or functional form cannot be determined. These difficulties can be found in the following two planes:

(1) Sudden changes of the condition connected with the randomness of external load intensity which can increase or decrease the wear rate of the object.
(2) Disturbances in the observation of the vibroacoustical symptom connected with the manner of the object's work or the measurement path.

For the purpose of condition prognosis, we must apply sophisticated methods taken from the analysis of time series [98] and econometrics [99]. However, the problem of prognosis in diagnostics is slightly different, since in each application we can assume that the technical condition of the object is deteriorating, sometimes even monotonically deteriorating. Therefore, it is something like the *prognosis of crisis or growth in econometrics.*

In the light of the above, our subsequent considerations about prognosis will be divided into two parts:

(1) Known or easy for determination condition–symptom model.
(2) Condition–symptom model unknown and almost impossible for analytical reconstruction.

4.5.2 Prognosis according to a known condition–symptom model

We assume that the relation of the disturbance amplitude to the value of symptom S is not great and that it is possible to reconstruct the analytical model of the symptom according to Equations (4.51)–(4.53) or from experimental data. The first possibility (i.e. the analytical reconstruction of the symptom) is not so easy at all and is not reliable, so that in the majority of cases we look for the model of symptom–trend changes on the basis of experimental data.

Assuming that the process of wear is a monotonic one, we can also look for monotonic representation into the vibroacoustical symptom. Taking into consideration only the simplest *models of symptom* trend, we can write successively a series of alternatives of the simplified function model (Equation (4.54)) in a general form:

$$S(\theta) = L[X(\theta)] + N(\theta) \tag{4.55}$$

We have the general formula and the resulting discrete forms for prediction purposes with the lifetime horizon $\Delta\theta$:

linear trend:

$$S_1 = S_0 + \lambda_1\theta + N(\theta) \rightarrow S_1[(n+1)\Delta\theta] = S_1(n\Delta\theta) + \lambda_1\Delta\theta$$

linear-square trend:

$$\begin{aligned} S_2 &= S_0 + \lambda_2\theta + \beta_2\theta^2 + N(\theta) \rightarrow S_2[(n+1)\Delta\theta] \\ &= S_2(n\Delta\theta) + \lambda_2\Delta\theta + \beta_2(2n+1)(\Delta\theta)^2 \end{aligned} \tag{4.56}$$

exponential trend:

$$S_3 = S_0 e^{\lambda_3 \theta} \cdot N(\theta) \rightarrow S_3[(n+1)\Delta\theta]$$
$$= S_3(n\Delta\theta)\, e^{\lambda_3 \theta}$$

square–exponential trend:

$$S_4 = S_0 e^{\lambda_4^2 \theta^2} \cdot N(\theta) \rightarrow S_4[(n+1)\Delta\theta]$$
$$= S_4(n\Delta\theta)\, e^{\lambda_4^2 (2n+1)(\Delta\theta)^2}$$

Here, S_0 is the initial value of the symptom, λ has the character of the intensity of changes and is proportional to the inverse of the failures time $\lambda \sim (1/\theta_{aw})$ or, better, $\lambda \sim (1/\theta_{zd})$ since the mean time of good condition $\theta_{zd} < \theta_{aw}$; β has the character of the acceleration of symptom changes and in the simplest case it can be $\beta \sim (1/\theta_{zd}^2)$ while $N(\theta)$ is a disturbance whose action should be taken into consideration when we estimate the values S_0, λ and β. This disturbance is the reason we must estimate the confidence intervals around the trend curve. The easiest way to do it is when we assume that $N(\theta)$ has the mean value zero and a constant variance. Then the determined trend can be treated as the current mean value, and the confidence intervals (e.g. 50%, 95%) can be determined according to known formula for the estimation of the mean confidence intervals [88, 97].

Observe that the differentiation of symptoms and some additional operations lead to a constant value disturbed only by noise $N(\theta)$. Hence, from the relation in Equation (4.56), we can obtain the indices of trend W_t in the following form.

(1) *Linear trend:*

$$W_{t_1} = \frac{dS_1}{d\theta} = \lambda_1 + n'(\theta) = \text{const} + n'(\theta), \quad n'(\theta) = \frac{d^2 N(\theta)}{d\theta}$$

(2) *Linear–square trend:*

$$W_{t_2} = \frac{d^2 S_2}{d\theta^2} = 2\beta_2 + n''(\theta) = \text{const} + n''(\theta), \quad n''(\theta) = \frac{d^2 N(\theta)}{d\theta^2}$$

(3) *Exponential trend:*

$$W_{t_3} = \frac{d \ln S_3}{d\theta} = \lambda_3 + n_e(\theta) = \text{const} + n_e(\theta), \quad n_e(\theta) = \frac{d \ln N(\theta)}{d\theta}$$

(4) *Square–exponential trend:*

$$W_{t_4} = \frac{d^2 \ln S_4}{d\theta^2} = 2\lambda_4 + n'_3(\theta) = \text{const} + n'_e(\theta), \quad n'_e(\theta) = \frac{d^2 \ln N(\theta)}{d(\theta)^2}$$

If we assume that the mean value of the corresponding derivatives of disturbance is zero, then the averaging of indices in Equation (4.57) by operator $E_\theta\,\{\mathbf{W}_t\}$ should give the indication of the trend of the real object. It will be that the model (4.56), for which the remainder after averaging of variance is minimal and in the range of stationarity $\theta > \theta_{stat}$, is not much changed with an increase of θ.

So, we already have the method of selecting the best-suited trend model with a simultaneous selection of the index of the speed of trend changes, since from Equation (4.57) we have, alternatively:

$$E_{\theta}\{W_{t_i}\} = \lambda, 2\lambda, 2\beta = \text{const}, \quad i = 1, 2, 3, 4 \tag{4.58}$$

So for a selected model k its variance is:

$$E_{\theta}\left[W_{t_k}(\theta) - E_{\theta}\left(W_{t_k}(\theta)\right)\right]^2 = \min$$

The initial value of symptom S_0 can be assumed as the first readout or the mean of the first two readouts. On the other hand, the lacking speed of changes of the square–linear model will be found as below:

$$\begin{aligned} &E_{\theta}\left\{\frac{d^2 s_2}{d\theta^2}\right\} \cong 2\beta \\ &E_{\theta}\left\{\frac{d s_2}{d\theta}\right\} - 2\beta\theta \cong \lambda \end{aligned} \tag{4.59}$$

which follows directly from the definitional relation of Equation (4.56) and from the averaging.

As an example of trend model determination for prognosis purposes let us take into consideration a fan type WPWD-55 in one of the power stations in Zielona Góra province*. At about 2-week intervals the horizontal peak accelerations of the external bearing of the fan unit were measured. The results of these measurements and the calculations determining the model type according to Equations (4.57)–(4.59) are presented in Table 4.2, where the derivatives were replaced, of course, by finite differences. The third column of the table gives an analysis of the linear model S_1; the fourth and fifth columns, the parabolic model S_2; the sixth column, the exponential model S_3; the seventh column the square–exponential model S_4. Analysing the results of this table, we see that the smallest value of the exponent, and particularly the most stable one with the smallest variance according to Equation (4.58) corresponds to the exponential model:

$$S_3(\theta) = S_0 e^{\lambda_3 \theta} \cdot N(\theta)$$

and then to the linear model:

$$S_1(\theta) = S_0 + \lambda_1 \cdot \theta + N(\theta)$$

---

* Data made available by courtesy of M. Zawiślak, MSc, a graduate of the Department of Postgraduate Studies of Diagnostics in Poznań Technical University in 1984.

Table 4.2 — Searching for a trend model of horizontal vibrations in the external bearing of fan WPWD-55

θ_k weeks	$S(\theta_k) = S_k$ m/s^2	$\frac{\Delta S}{\Delta \theta} = \lambda_1$	$\frac{\Delta^2 S}{\Delta \theta^2} = 2\beta_2$	$E_0\left[\frac{\Delta S}{\Delta \theta}\right] - \theta_k \cdot E_\theta\left[\frac{\Delta^2 S}{\Delta \theta^2}\right] = \lambda_2$	$\frac{\Delta(\ln S)}{\Delta \theta} = \lambda_3$	$\frac{\Delta^2(\ln S)}{\Delta \theta^2} = 2\lambda_4$	Remarks
0	36	—	—	—	—	—	The best model:
2	31	-2.5	—	1.54	-0.075	—	$S_3 = S_0 \cdot e^{\lambda_3 \theta} \cdot N(\theta)$
4	39	4.0	3.25	1.62	0.115	0.0948	Prognosis* of succesive
6	42	1.5	-1.25	-4.78	0.037	-0.0389	symptom values =
8	44	1.0	-0.25	-7.94	0.023	-0.0069	$S_3(n+1) = S_n e^{\lambda_3 \Delta\theta}$
10	47	1.5	0.25	-11.10	0.033	-0.0049	e.g.:
12	50	1.5	0.00	-14.26	0.031	-0.0010	$S_3(18) = 102$ m/s^2
14	72	11.0	13.25	-17.42	0.182	0.0757	instead of 112 m/s^2
16	90	9.0	-1.00	-20.58	0.112	-0.0354	$S_3(20) = 127$ m/s^2
18	112	11.0	1.00	-23.14	0.109	-0.0011	instead of 130 m/s^2
20	130	9.0	-1.00	-26.90	0.075	-0.0174	$S_3^*(22) = 130 \cdot e^{2 \cdot 0.064}$
22	147	—	—	—	—	—	$= 147.7$m/s^2
$E_\theta(\cdot)$	63	$\underline{\underline{47 > 0}}$	$1.55 > 0 \neq$ const	$-12.63 < 0$	$0.064 > 0 =$ const	$0.0072 > 0 \neq$ const	

Note: $E_\theta = \frac{1}{N} \sum_{k=1}^{N} (\cdot)$ $\Delta^2 S = S_k - 2S_{k-1} + S_{k-2}$

$\Delta S = S_k - S_{k-1}$

The resulting differential formula for the prediction of value $S\,[(n+1)\,\Delta\,\theta] \equiv S\,(n+1)$ with known $S\,(n)$ are:

$$S_1(n+1) = S_1(n) + \lambda_1 \Delta\theta$$

$$S_3(n+1) = S_3(n)\, e^{\lambda_3 \Delta\theta}$$

On this basis, the forecast value of the twentieth week on the basis of the eighteenth one we have ($\Delta\theta = 2$ weeks):

$$S_1(20) = S_1(18) + 4.7 \cdot 2 = 121.4\ \mathrm{m/s^2} \quad \text{instead of the measured130 m/s}^2$$

$$S_3(20) = S_3(18)\, e^{0.064 \cdot 2} = 127\ \mathrm{m/s} \quad \text{instead of the measured130 m/s}^2$$

Therefore, we can accept with great certainty that, in this case, the exponential trend will be best fitted and the successive values of the acceleration of bearing vibrations will be found from the relation:

$$S_3[(n+1)\,\Delta\theta] = S_3(n\,\Delta\theta)\, e^{0.064 \cdot \Delta\theta}$$

and the value not yet measured in time $\theta = 22$ weeks will be:

$$\underline{S_3(22\ \text{weeks}) = 147.7\ \mathrm{m/s^2}}$$

Terminating the discussion on the problem of prognosis according to a known trend model, it is useful to illustrate this model in a graphical way, as in Fig. 4.15 where we see the question of the determinism of the trend in the form of the realized continuous curve and the unrealized broken part of it. Deviations of the measured symptom values are treated here as disturbances or machine observations Having obtained from the calculations the best adjustment of trend, we can estimate the residual time of utilization θ_r at any time of the current observation of the symptom.

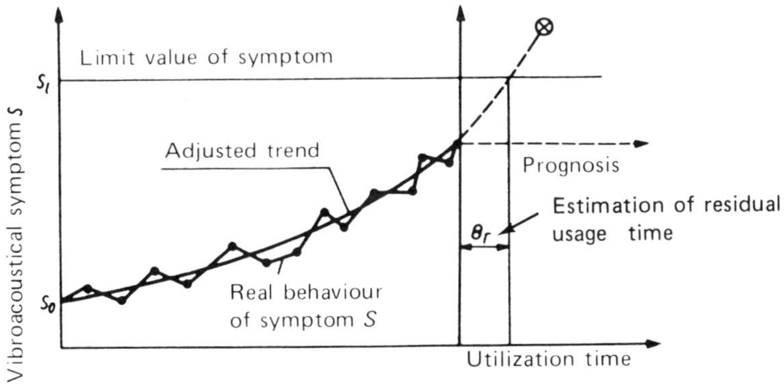

Fig. 4.15 — Illustration of the prognosis of the residual utilization time according to a known trend model

4.5.3 Condition prognosis with an unknown trend model

Although, in many cases, the prognosis according to a known trend model, as in Section 4.5.2 stands the test in diagnostics, there exist some difficult cases with high levels of disturbances which do not obey this rule. If, in addition, they are unitary objects with a high responsibility where statistics cannot be used, then the only solution is the current construction and improvement of the trend model on the basis of the successive inflowing data about the symptom values $S(\theta_k)$. The most important methods giving such possibilities comprise: (a) Brown's exponential equalization [99, Chapter 6; 100]; (b) autoregression technique; (c) technique of the moving average; and (d) the mixed techniques of autoregression moving average and autoregression integral moving average [98, Chapters 3 and 4]. In diagnostics we do not have any long series of observed symptom and the mean number of observations oscillates between ten and twenty. As argued by Batko [100] in such cases the priority belongs to Brown's exponential smoothing. Therefore, we will limit our considerations only to the explanation of the ideas of the remaining condition prediction techniques.

Let us assume that the successive values of the observed vibroacoustical symptom are $S(\theta_k) = S_k$; $k = 1, \ldots, l$. According to Brown, the estimation of trend T_k of such a series on its k-th step can be expressed:

$$\begin{aligned} T_k &= \alpha S_k + (1-\alpha) T_{k-1}, \quad k = 2, \ldots, l \\ T_0 &= 0, \quad T_1 = S_1 \end{aligned} \tag{4.60}$$

Knowing this evaluation we can calculate the prognosis, outpacing it by one time segment, i.e. for $\theta_{l+1} = \theta_l + \Delta\theta$ counted in place l:

$$\begin{aligned} S_p(l+1) &= T_1 + (T_1 - T_{l-1}) \frac{\Delta\theta}{\theta_i - \theta_{i-1}}, \quad l = 2, 3, \ldots \\ S_p(1) &\equiv S_1 \\ S_p(2) &\equiv S_2 \end{aligned} \tag{4.61}$$

Now, if we treat the prognosis $S_p(l+1)$ as a new readout of the value of symptom S_{l+1}, then in this way we can extend the prognosis in place $\theta = \theta_k$, but only with the possibility of greater error. As we see, it is a very simple technique of prognosis with one unknown called the *equalization constant* $0 < \alpha < 1$. The selection of this constant is not algorithmed in a simple way, and its value is generally selected by guessing while taking care that the mean error of prognosis Δs are small. Here, the following formulae are necessary:

mean error of prognosis

$$\bar{\Delta} S = \frac{1}{L} \sum_{k=1}^{L} \left[S_p(k) - S_k \right]$$

standard deviation of error

$$\sigma_{\Delta s} = \sqrt{\frac{1}{L} \sum_{k=1}^{L} \left[S_p(k) - S_k \right]^2} \tag{4.62}$$

where L is the number of pairs $(S_p(k), S_k)$ (i.e. the previous prognosis and the value observed later) accepted for the verification of prognosis. With these data it would be necessary to evaluate the confidence interval of prognosis; unfortunately, available sources [98, 99] do not discuss this problem in relation to Brown's method.

For an illustration of this technique for improving the trend function and for the prediction of a symptom by way of exponential equalization, let us consider the same data of fan WPWD-55 as in Table 4.2 and compare them with Table 4.3. Table 4.3

presents successive time moments of observation θ_k; values of symptom S_k; and for the successive values $\alpha = 0.3$, 0.5 and 0.8 the values of trend estimation T_k (Equation (4.60)); prognosis $S_p\ (k+1)$ (Equation (4.61)) and prognosis error $S_p\ (k+1) - S_{k+1}$ and its square. The last row of the table presents the mean values $E_\theta\ (\cdot)$ or their roots from the mean square. Looking from this angle on the minimum of mean deviation from prognosis and the minimal value of the mean square of deviation, we find that for our case of the prediction of fan vibrations, the best value of the equalization constant $\alpha = 0.5$. It gives minimal mean deviation $E_\theta\ [S_p\ (k) - S_k] = -4.62$ and standard deviation

$$\sqrt{E_\theta\left[S_p(k) - S_k\right]^2} = 4.61$$

underlined with double lines in the table.

So the best prediction of the acceleration value of the future measurement for $\theta = 22$ weeks will be $S_{22} = 148.3\ \text{m/s}^2$. Comparing this value with the prediction for the exponential model where $S_3\ (22) = 147.7\ \text{m/s}^2$, we find almost complete agreement of both values. So, we see that both methods of prediction according to the known model and Brown's adaptation method can be applied in this case with the same good results.

Summing up the prognosis by Brown's method, we can say that it depends on the regularity of the selection of the equalization constant α. Although Batko [100] argues that this selection can be algorithmed, it is not so simple; it is just as complicated as the autoregression moving average methods.

The autoregressive model of symptom trend assumes that each observation is a weighted and finite sum of the previous ones plus an independent random factor [99, Chapters 3 and 4]. So the prognosis $S_p\ (k)$ outpaced by one $\Delta\theta$ (for simplicity's sake) can be written as:

$$S_p(k) = \psi_1 S_{k-1} + \psi_2 S_{k-2} + \ldots + \psi_r S_{k-r} + b_k \tag{4.63}$$

where r is the number of the previous observations taken into consideration; ψ denotes coefficients of the weight of observation determined with the help of special procedures; and b_k is the random component (impulse) of white noise type.

In the model of the trend of the moving average type it is assumed that each observation is a reaction of an unknown linear system to white noise. In the finite discrete version, a one-stage prognosis can be written as:

$$S_p(k) = b_k - \upsilon_1 b_{k-1} - \upsilon_2 b_{k-2} - \ldots - \upsilon_q b_{k-q} \tag{4.64}$$

where b denotes, as previously, random impulses of the white noise type and υ denotes the discrete coefficient of the weight of the system whose entrance is influenced by b. These coefficients, similarly as for Equation (4.63) can be determined with the help of special procedures given, for example, in the fifth part of the monograph by Box and Jenkins [98].

The combination of the properties of both models gives the autoregressive moving average (ARMA) procedure while, with a small modification consisting of the summing (integration–integral–(I)) on a small segment of initial series, we will obtain the procedure of trend reconstruction called autoregressive integral moving average (ARIMA). The formulae and procedures of autoregressive integral moving average together with the confidence intervals are given by Box and Jenkins [98, Chapter 5]. Therefore we will not discuss them here, but it is worth illustrating graphically (Fig. 4.16) the type of prognosis possible in this case.

Table 4.3 — Prediction of the trend of fan bearing vibration acceleration according to the method of exponential equalization for the values of equalization factor α

k	θ_k	$S(\theta_k) = S_k$	$\alpha = 0.3$				$\alpha = 0.5$				$\alpha = 0.8$			
	weeks	m/s²	T_k	$S_p(k)$	$S_p(k)-S_k$	$(S_p(k)-S_k)^2$	T_k	$S_p(k)$	$S_p(k)-S_k$	$(S_p(k)-S_k)^2$	T_k	$S_p(k)$	$S_p(k)-S_k$	$(S_p(k)-S_k)^2$
1	0	36	36.00	36.00	0.00	0.0000	36.00	36.00	0.00	0.000	36.00	36.00	0.00	0.00
2	2	31	34.50	31.00	0.00	0.0000	33.50	31.00	0.00	0.000	32.00	31.00	0.00	0.00
3	4	39	37.02	31.00	-8.00	64.0000	36.25	28.50	-10.50	110.250	37.60	24.00	-15.00	225.00
4	6	42	38.50	42.04	0.04	0.0016	39.16	41.74	-0.25	0.626	41.12	48.80	6.80	6.24
5	8	44	40.15	41.50	-2.50	6.2500	41.58	44.98	0.90	0.810	43.42	48.16	4.16	17.31
6	10	47	42.20	43.45	-3.55	12.6000	44.29	46.42	-0.58	0.340	46.28	48.02	1.02	1.04
7	12	50	44.54	46.31	-3.69	13.6200	47.15	49.71	-0.29	0.840	49.26	52.00	2.00	4.00
8	14	72	52.78	49.22	-22.78	518.9300	59.58	52.87	-19.13	365.960	67.45	55.22	-17.78	281.57
9	16	90	63.95	69.26	-20.74	430.1500	74.79	84.44	-5.56	30.910	85.49	103.83	13.83	191.27
10	18	112	78.37	86.29	-25.71	661.0000	93.40	105.21	-6.76	46.100	106.70	121.57	9.57	91.58
11	20	130	93.86	107.21	-22.79	519.4000	111.70	130.62	0.62	0.380	125.34	149.12	19.12	365.57
Prognosis	22	—	—	124.84	—	—	—	148.30	—	—	—	162.62	—	—
$E_\theta(\cdot)$	⇒	63	51.08	59.01	12.19	$\sqrt{E_\theta} = 15.73$	56.13	66.65	-4.62	$\sqrt{E_\theta} = 4.61$	60.00	73.36	2.64	$\sqrt{E_\theta} = 11.66$

Note: $T_k = \alpha S_k + (1-\alpha)T_{k-1}, \quad k > 1 \qquad S_p(k+1) = T_k + \left(T_k - T_{k-1}\right)\Delta\theta, \quad k > 2 \qquad E_\theta(\cdot) = \frac{1}{N}\sum_{1}^{N}(\cdot)_k$

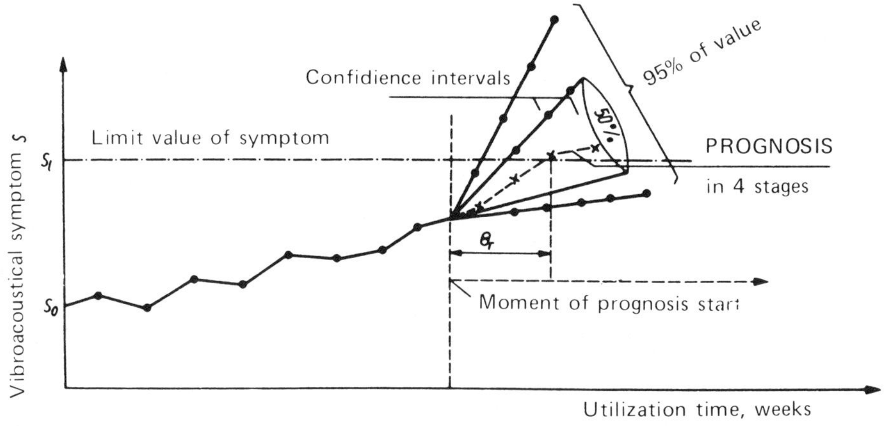

Fig. 4.16 — Qualitative illustration of the prognosis according to the model of the autoregressive integral moving average trend according to Box and Jenkins [98] and the estimation of the residual utilization time θ_r

Such a prognosis could be carried out by a computer supervising the condition of unitary machines and assemblies with a high responsibility, e.g. power units. Then it pays to develop a software package of prognostic programs or even to program them in the read-only-memory of a computerized supervision system so that in each point of utilization time the estimation of residual utilization time is known with a preset level of confidence.

4.6 SUMMARY

The important problem of diagnostic inference stresses the following important problems connected with the actual and future estimation of the condition of a mechanical object.

The problem of first importance is the knowledge of the diagnostic model assigning the relation symptom $\rightarrow$ condition. Looking at the difficulties connected with the inference with the probabilistic diagnostic model, it is necessary to take care to obtain a pseudodeterministic diagnostic model. As shown in Chapter 3, the possibility of finding the trend, or the life curve, exists even with an unknown trend model. This reasoning is correct not only for an object with a one-dimensional domain of failure, but also can be repeated for each dimension (symptom) separately, or reducing the problem to one-dimension by the creation of a cumulant. The question of finding the limit value—one for the whole mechanical object—can be clarified, of course, if such simplification is allowed in the optimal management of the object's utilization.

5

Diagnostics case histories in industry

5.1 INTRODUCTION

At this point, we can say that we have analysed all elements from the intention to implement diagnostics, to taking an objective utilization decision about a machine. The user, therefore, may frequently ask the question: Well, vibroacoustical diagnostics is interesting and encouraging for application, but how am I really to start applying it? This chapter is an answer to this question and presents a general course of procedure, with definite examples and problems for implementing vibroacoustical diagnostics in industry.

5.2 GENERAL SCHEME FOR THE IMPLEMENTATION OF VIBROACOUSTICAL DIAGNOSTICS

In a particular industrial plant, the utilization and reliability characteristics of specific machine groups are the essential factors conditioning the final form of the diagnostic system in the plant. By the diagnostic system we will understand here a system consisting of the following.

(1) Diagnosis of objects (machines, etc.).
(2) Diagnostic means (instrumentation and personnel, if the system is not automatically operated).
(3) Procedures for diagnosis which ensure a method of mutual interaction to fulfil the tasks of an objective cognition of the technical condition or the quality of a mechanical object.

It follows that our general procedure in the implementation of diagnostics must take into consideration the determination of the following three diagnostic components.

(1) Object or objects being diagnosed, i.e. machines and installations which are to be the subject of diagnosis.
(2) Diagnostic means, i.e. instrumentation, personnel, computers, permitting the gathering and processing of vibroacoustical signals.
(3) Diagnostic procedures defining the manner of collection, (e.g. a continuous remote measurement, or periodic measurement) and processing the signals and diagnostic information (by man or by microprocessor, computer, etc.).

Usually, the determination of these three elements of the diagnostic system can follow different paths, and the following five-point system of approaching the problem of determination should be treated as one of the possible ways. It is presented as a working method of approach used in a recent postgraduate study in diagnostics carried out by the Author and his research team.

(A) Selection of the objective and the object of diagnostics

Many people believe that knowledge about the objective and the object of diagnostics is accessible to everybody. Except for the case of a single simple machine, in the plant the problem is not so simple. If we express the final function of the objective of diagnostics in the categories of financial means to be gained or lost in the given plant, then the choice of a machine group and/or subassembly in a machine selected for diagnosis is not so obvious. Therefore, for proper determination of the objective and object of diagnosis we should carry out *reliability and economic analyses of the machine stock* in the plant, and the same analyses should be carried out for the selection of subassemblies/elements of a machine requiring objective evaluation of its technical condition.

(B) Selection of place and measurement magnitude

Having a definite type of machine and its subassemblies which require diagnostic estimation of condition, we should analyse the degradation processes which cause failures. This analysis should consider particularly the vibroacoustical processes accompanying the given type of wear and the ways of its propagation in the machine construction. We will define, thereby, the dimensions of the failure space of the machine, their possible interrelations and the type of measurement magnitude interchangeable with the given degradation process and the place of its measurement with the smallest level of disturbances. For example, looking for vibroacoustical measurement values interchangeable with the condition of the blade ring of a turbomachine, we will select the measurement of vibration velocity on the machine body, separating it in this way from the high-frequency signals of cavitation and bearings, as well as gaining a spatial distance from the signals of failures connected with the drive (unbalance, misalignment).

(C) Selection of measures representing vibroacoustically the life curves of failure

In order to carry out this task, we should first decide which variant of a passive experiment will be taken into consideration. In the case of a simultaneous observation of many objects in different lifetimes (conditions) or successive observations of one object, we should make a recording of vibroacoustical signals for further laboratory treatment or carry out as many local measurements of vibroacoustical symptoms as possible. In the laboratory the recorded signals should be transformed into symptoms using, for their initial selection, the knowledge contained in Chapter 3. For further diagnostic inference, we select a number of symptoms not smaller than the number of the supervised failures.

These symptoms should be possibly independent, and represent the dynamics of the life curve. By a good representation we understand here the monotonically growing curve $S(\theta)$ in the lifetime θ.

In the distribution field of the amplitudes of symptom $p(S)$ the representation should be positively (right-hand skew) asymmetrical with a maximum around the initial value S_0, as in Fig. 4.1 or Fig. 4.6.

The selection of vibroacoustical symptoms representing the life curve can be significantly facilitated, or even computerized, by the application of the methods of discrimination, orthogonalization and decomposition of time series. An example of the operation of such methods consisting of the selection of failure oriented symptoms called BEDIND (passive experiment) will be discussed in Chapter 6.

(D) Determination of the relation symptom → condition

For selected vibroacoustical symptoms of the condition the limit values of the GOOD CONDITION S_l should be determined with the help of the existing and verified standards or the statistical decision theory discussed in Section 4.3. The determined limit values should be corrected from time to time according to the actual observations of the utilized machines.

(E) Organization of diagnostics and utilization decisions

The methods of obtaining, processing and storing the signals, symptoms and other diagnostic information should be defined instrumentally and operationally (instruments and personnel). On the other hand, a similar precision should be applied to the determination of the way of taking, transferring and carrying out the utilization decisions in the complex structure of utilization management in the plant.

The execution of the above five points will ensure the creation of a diagnostic system with a high level of reliability in diagnostic decisions and a high operation effectiveness.

5.3 DIAGNOSTICS OF HIGH-POWER FANS IN POWER ENGINEERING

In the Patnów-Adamów-Konin power plant complex an attempt has been made to base the overhauls of the auxiliary installation of the power unit on the results of vibration diagnostics as detailed in Krugiełka [93].

In trying to answer question (A) at the conclusion of Section 5.2 above, referring to the objective and the first-plan object of diagnostics, we have reached for the available data about the annual time of breakdown standstill (Fig. 5.1) and the annual costs of repairs (Fig. 5.2) of machines and auxiliary installations of the power unit. The comparison of the premises for the decision from both figures does not encourage any unequivocal decision. However, selecting the yearly costs of overhaul as being the more essential ones, in the further part of the work [93] Krugiełka has dealt with the exhaust and blowing fans. Departing from the premise of the physical generation of vibrations connected with mechanical and aerodynamical unbalance and misalignment, a decision was taken to measure the vibration velocity, while the field investigations were to answer the remaining questions.

The results of four-month investigations into the vibrations of the exhaust fan

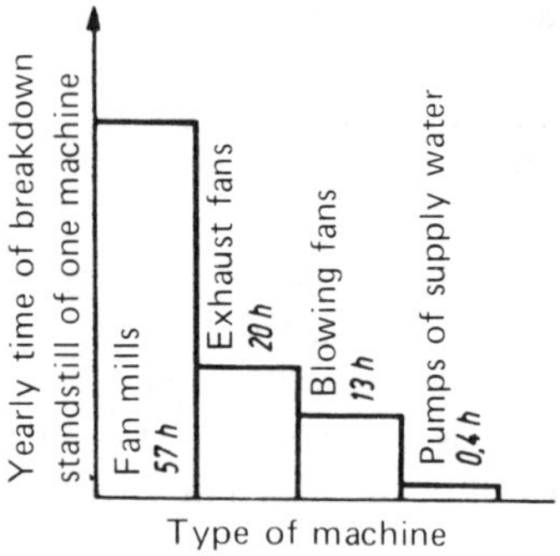

Fig. 5.1 — Comparison of the yearly standstill time for different auxiliary machines in a power unit [93]

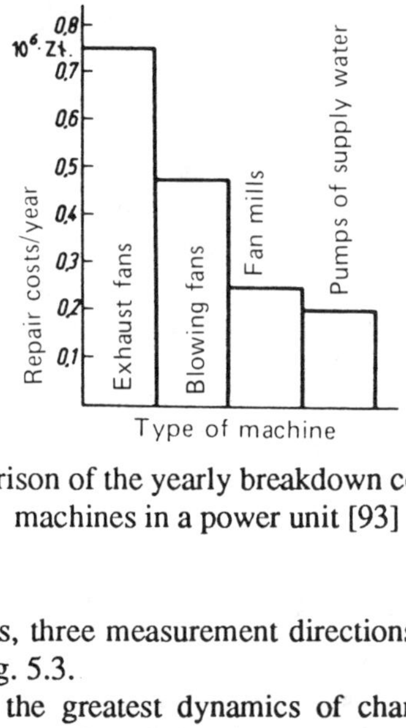

Fig. 5.2 — Comparison of the yearly breakdown costs of the auxiliary machines in a power unit [93]

BAB - 106, for two bearings, three measurement directions and two velocity amplitude measures are presented in Fig. 5.3.

As we see in Fig. 5.3, the greatest dynamics of change are characteristic of the horizontal and axial vibrations of both bearings measured in the categories of peak velocity. This fact has been confirmed in other fans. Therefore, one can recommend the measurement of both bearings measured in the categories of peak velocity in the horizontal H and axial A directions.

Taking measurements of these fans in four power units, data have been obtained for the determination of the limit value of the peak vibration velocity according to the methods based on the Neyman-Pearson formula, according to Section 4.3.3. Figures 5.4(a) and 5.4(b) present results of these investigations and calculations showing ~ 3% and ~ 4%, respectively, of unneeded overhauls using an inference rule elaborated in this way with an availability $P_z = 0.9$ and the admissible level of unneeded overhauls A = 10%. These values should be verified when we undertake the routine application of the diagnostic evaluation of auxiliary installations condition.

As the final example of the implementation of vibroacoustical diagnostics in high-

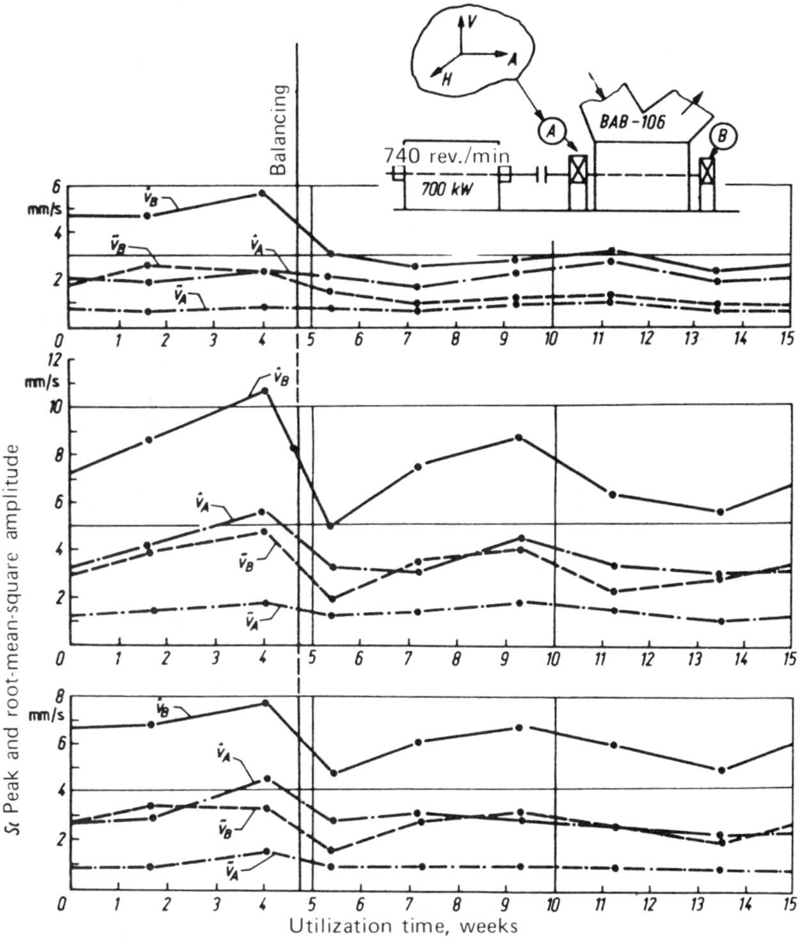

Fig. 5.3 — Utilization trend of the vibration velocity of the bearings of fan BAB - 106 No. IV (slide bearing) [93]

power fans let us take into consideration WPWD-55 fans with rolling element bearings.* Here, during 15 weeks, only the accelerations of two bearings were measured in three directions. The greatest changes were shown by the vertical acceleration measurements shown in Fig. 5.5. This figure indicates that the best representation of the condition (life curve) is provided by the measurement of peak accelerations as well as the impulse factor $I = \hat{a} / \bar{a}$ or the peak factor $C = \hat{a} / a_{rms}$; these values are not shown in the figure but are easily calculated on the basis of the presented data.

Finally, as it seems from the above, the moments of the implementation of vibro-acoustical diagnostics of high-power fans in power engineering have been sufficiently illustrated. Problems in the general procedure will certainly become easier to solve by analysing other examples.

---

* Data for the example were made available by the courtesy of M. Zaswiślask, MSc.

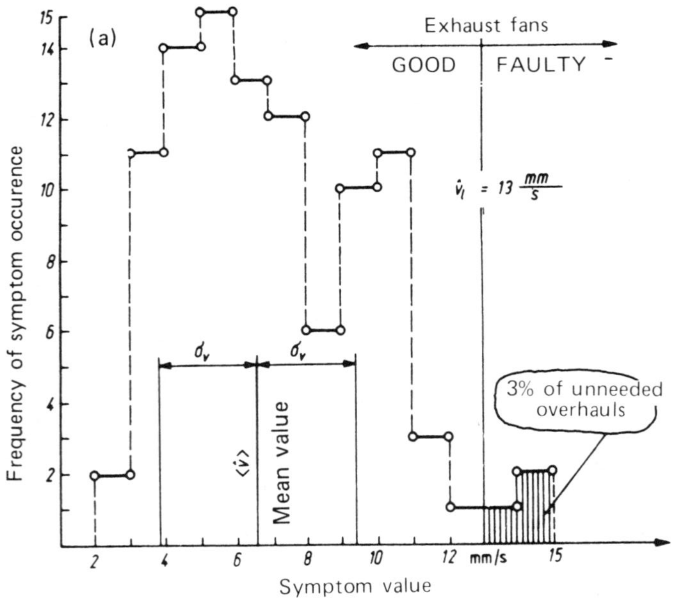

Fig. 5.4(a) — Determination of the limit value of the peak velocity of horizontal vibrations *H* of the bearing A of BAB-106 fan [93]

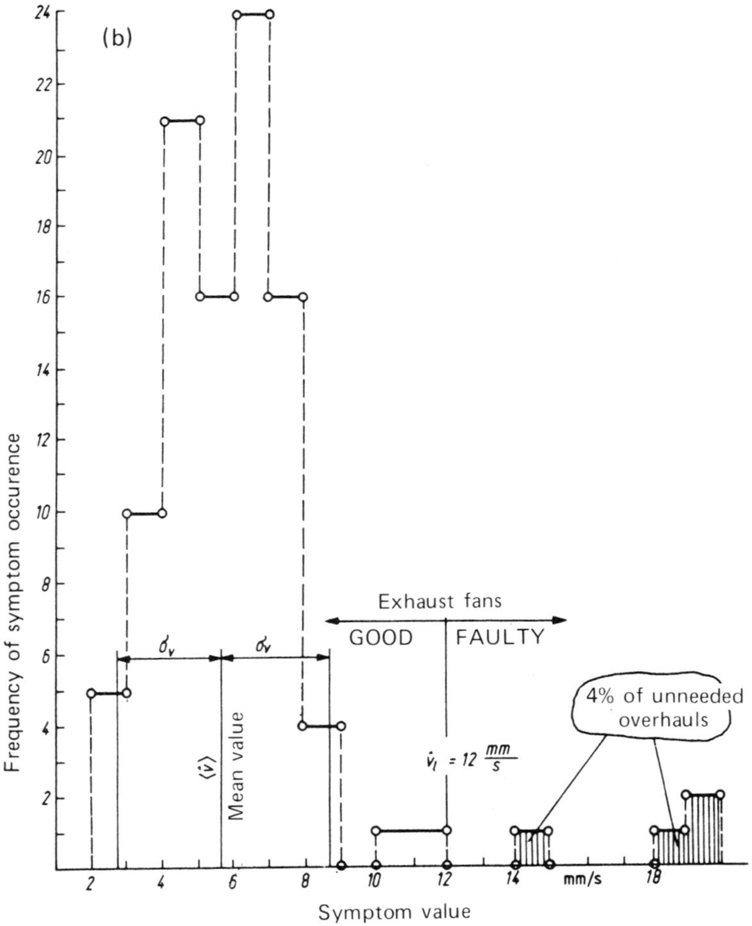

Fig. 5.4(b) — Determination of the limit value of the peak velocity of axial vibrations *A* of the bearing B of BAB-106 fan [93]

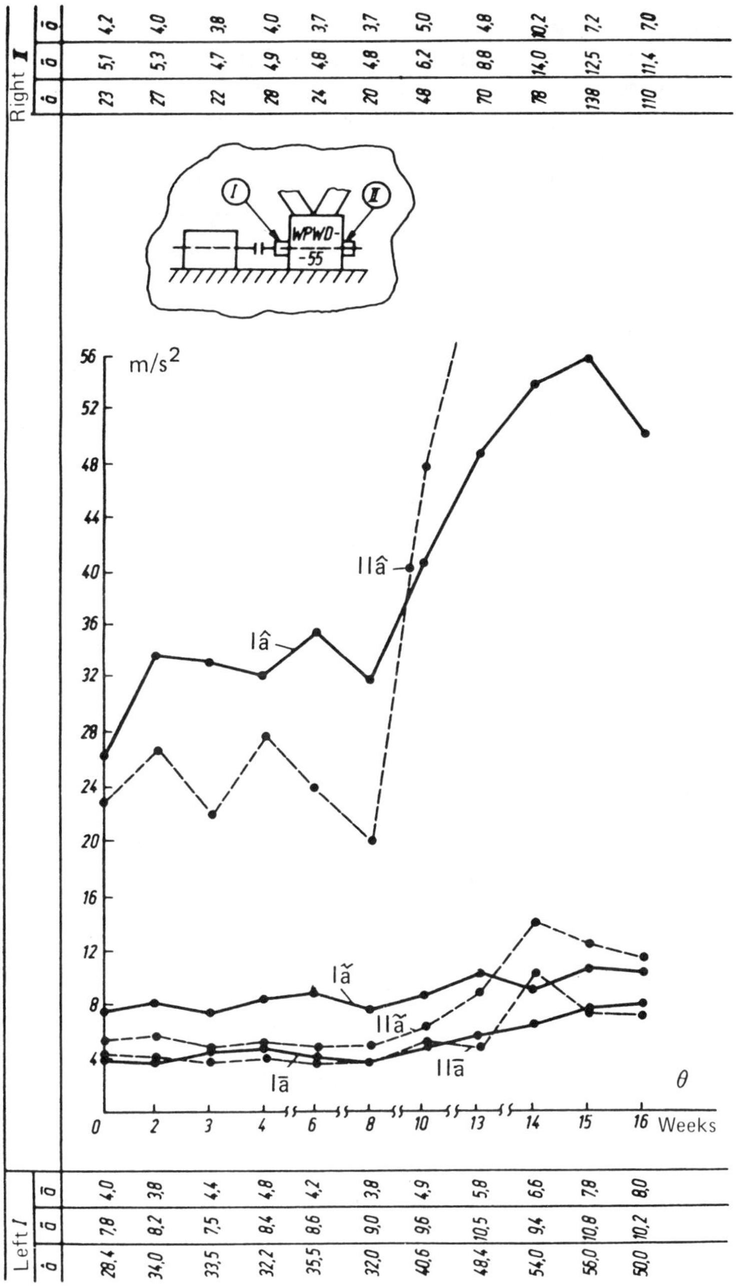

Fig. 5.5 — Trend of changes in acceleration of vertical bearings of WPWD-55 fan

5.4 DIAGNOSTICS OF DIESEL TRACTION ENGINES

A significant part of the drive of railway rolling stock consists of locomotives with diesel engines. Preventive planned repairs lead to situations in which elements are replaced which are still good for further use. For example, Tomaszewski [101] stated that 53% of elements of the locomotive ST-43 did not exceed the limit values of the GOOD CONDITION and in spite of that they were replaced according to the procedure of repairs. Therefore, the constitution of an objective checkout method of the GOOD CONDITION, (e.g. by the vibroacoustical diagnostics) is an urgent reliability and economic problem of the transportation system. Accepting the repair costs as the decision index, Tomaszewski has analysed [101] the relative costs of the particular subassemblies of the SF-45 locomotives and, as can be seen from Fig. 5.6, electric machines and internal combustion engine are demonstrated to be primary objects for diagnostics.

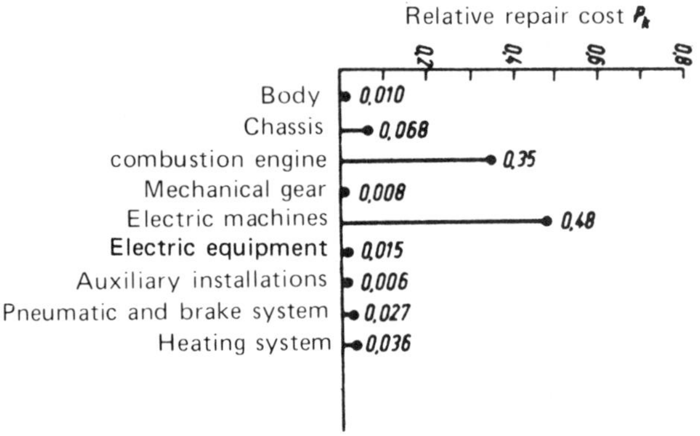

Fig. 5.6 — Histogram of the participation of repair costs of the particular assemblies of the SP-45 locomotive [101]

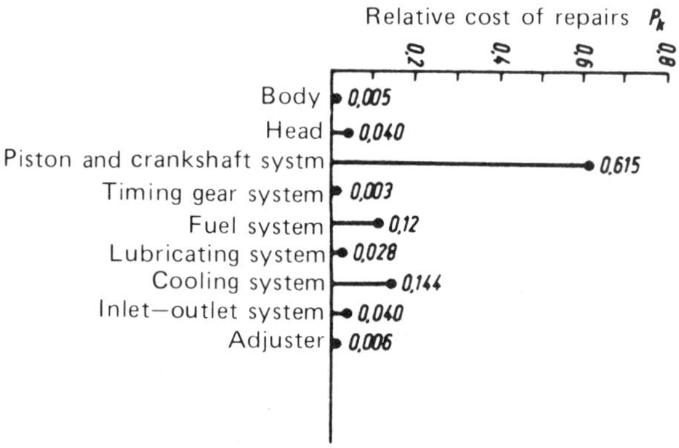

Fig. 5.7 — Histogram of the participation of repair costs of the particular systems of the SP-45 locomotive combustion engine 2112SSF [101]

Referring to the engine, the greatest repair costs are due to the piston–crankshaft system (61.5%) which is clearly seen from Fig. 5.7. Thus, the measurement points of vibrations and noise of a locomotive diesel engine are localized on the line of the crankshaft bearing and on the line of the upper dead point *Z–Z* of the pistons (Fig. 5.8). Tomaszewski [102] shows the trend of changes in the velocity and vibration acceleration in the middle point of the *Z–Z* line (Fig. 5.9) as a function of the mileage of the locomotive.

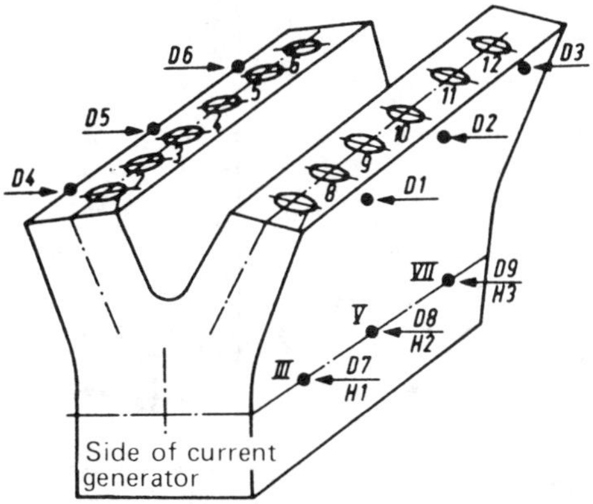

Fig. 5.8 — Place of vibration signal (D) and noise signal (H) reception on the body of the locomotive engine SP-45 [101]

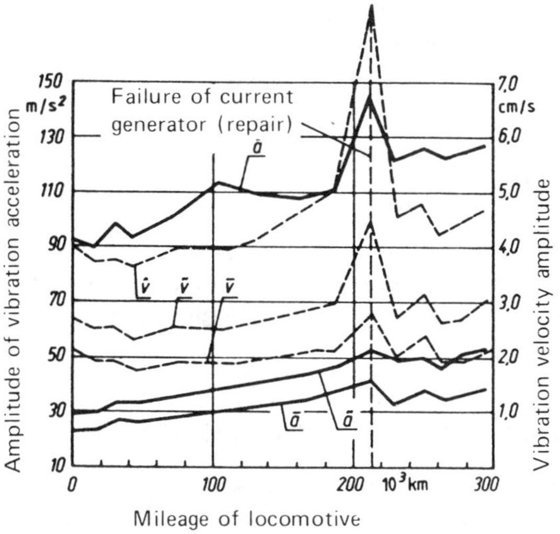

Fig. 5.9 — Changes in the amplitude of acceleration as a function of the mileage of the locomotive in point D2 where $\hat{a}$ is peak acceleration, $\tilde{a}$ is root-mean-square acceleration, $\bar{a}$ is mean acceleration and the course of changes in the amplitude of vibration velocity in point D2, a visible increase corresponds to the damage of the electric current generator [101]

The greatest sensitivity to a gradual degradation of the engine and a sudden damage of the generator are characterized by the peak values of acceleration and particularly velocity. Similar properties are characteristic of the peak coefficient C and the impulse factor I of these values which can be easily deduced from Fig. 5.9. Thus, a gross selection of measurement value can be regarded as a satisfactory one v or I_v. There still remains the question of the localization of the failures and the determination of limit values of the measured symptoms.

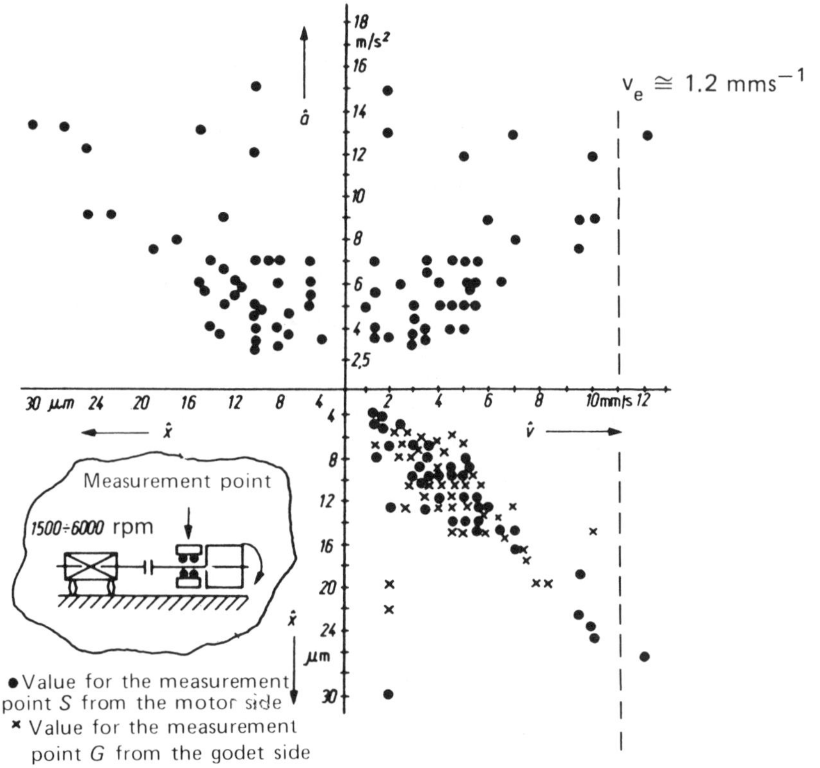

Fig. 5.10 — Qualitative illustration of the interdependence between the displacements, velocities and accelerations of vibrations in a group of forty godets [103] and the scheme of vibration measurement

5.5 DIAGNOSTICS FOR TEXTILE MACHINES

In the formation and winding of silk yarn, assemblies are used consisting of an electric motor (with rotations controlled by the frequency of the supply current) and a heated spindle on which the yarn is wound. The spindle on a rolling element bearing is connected with the motor by an elastic clutch and it rotates at speeds of 1500–6000 rev./min. The basic failures of this assembly, which is called a godet, include the wear of rolling element bearings and unbalance or bending of the shaft since the spindle is pivoted in the bearing assembly [103]. In looking for the diagnostic value, the peak values of acceleration $\hat{a}$, velocity $\hat{v}$ and of displacement $\hat{x}$ of the radial vibrations of the bearing assembly from the motor side were measured. By investigating 160 godets, it was found that the readouts of accelerations and displacements were almost independent, which proves that different diagnostic information was transferred (Fig. 5.10). Therefore, the measurement of displacements $\hat{x}$ for the detection of unbalance and shaft bending, and the measurement of accelerations $\hat{a}$ for the evaluation of the condition of bearings, were postulated. Evaluating the availability index of godets for P_z = 0.75 =75% and the admissible probability of unneeded overhauls A = 0.1 = 10% from a passive experiment, the following limits of the Good Condition of godets were obtained: $\hat{a}_l = 21.2$ m/s^2; $\hat{v}_l = 11.9$ mm/s; $\hat{x}_l = 36.3$ μm. Thereby, we have obtained data for the introduction of the vibrational evaluation of the technical condition instead of the planned preventive overhauls.

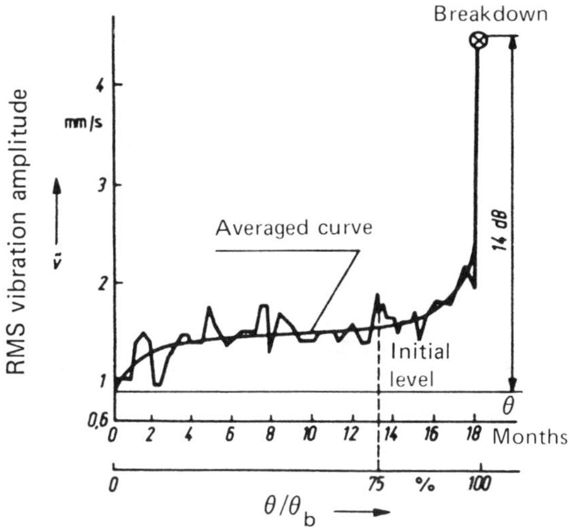

Fig. 5.11 — Changes of the root-mean-square velocity of vertical vibration of bearing casing SE [104]

5.6 DIAGNOSTICS OF ELECTRIC MOTORS

Electric motors of low power are frequently used to drive auxiliary installations. The Author [7] considers the defects which can occur here during use, are, for example, wear of roller bearings, loosening of wiring, jamming of rotor, and friction of fan blades, etc. In the diagnostic investigations of such drive sources, a motor was considered with the power of 0.22 kW, speed $n = 1375$ rev./min, and bearings 6203Z, and its degradation was observed in motion under load. In the investigations we have measured the acceleration of vertical vibrations of the internal bearing casing recording it for a later analysis [104]. The degradation of the technical condition of the motor represented by the velocity of vibrations is shown in Fig. 5.11 and vibration acceleration in Fig. 5.12.

The dynamics of the velocity changes from start to breakdown is 14 dB, while in the accelerations we obtain 31 dB. It means that the failure generates vibrations of high frequency. In the acceleration band of 20 Hz–2 kHz, an increase of 15 dB was recorded. It has indicated definitely that the bearing was the cause of failure, which was subsequently confirmed. So we can see that for the observation of the failure of rolling element bearings in electric motors we should observe the vibration acceleration.

5.7 SUMMARY

Summing up this chapter dealing with practical application, we must recommend other

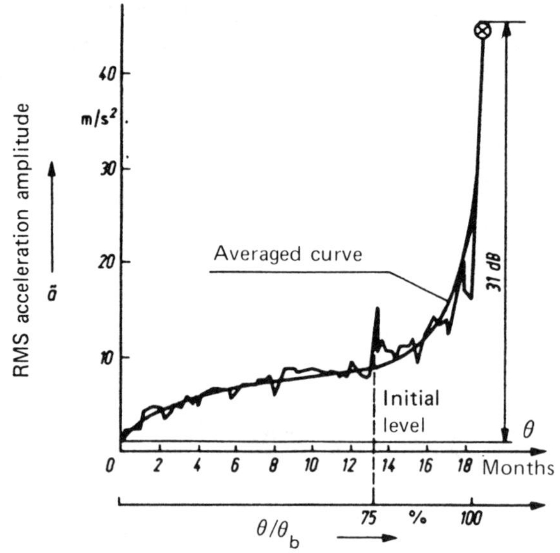

Fig. 5.12 — Changes of the root-mean-square value of the acceleration of the vertical vibration of bearing casing SE [104]

examples of diagnostic implementation described in the literature. There are many monographs [7] quoted already in the works referring to the diagnostics of a pumping engine [72] and the high-power engine [73]. Readers interested in the diagnostic of rolling element bearings will find plenty of case histories in the literature [105, 106].

Readers interested in the diagnostics of vehicles also can find material [107].

6

Advanced problems of vibroacoustic diagnostics

5.1 INTRODUCTION

The theory and application examples so far described will be enough for the majority of readers employed in industry (i.e. for engineers working in diagnostic services or quality inspection) to preform their duties efficiently. However, in vibroacoustical diagnostics not all is clearly defined and carried out in the optimal way. The best proof of that is detailed elsewhere by the Author [83] where the state of vibroacoustical diagnostics and the problems requiring particular attention are described. Parts of the problem are in the stage of the initial application, and it seems advisable to present their nature and possible utilization values so that the more advanced and creative users of vibroacoustical diagnostics can utilize it to increase the objective evaluation of the technical condition of machines. These possibilities are contained in the organization of a diagnostic experiment (Section 6.2), in the handling of its results, and in the processing of diagnostic information which will be discussed in Sections 6.3 and 6.4, as well as in another approach to the observation method of diagnostic signals of mechanical objects, which will be reported in Section 6.5.

6.2 DIAGNOSTIC EXPERIMENTS

Żółtowski [108] states in his dissertation that 'a diagnostic experiment is an organized action in order to obtain information about the process or object in which we are interested'. In diagnostics we are interested in the effects of the degradation of the utilization properties of mechanical objects. Therefore, the objectives of a diagnostic experiment are as follows.

(1) Discrimination of vibroacoustical characteristics (symptoms) of the condition and quality.

(2) Recognition of the condition or quality.
(3) Discrimination of vibroacoustical signals being the best carriers of this diagnostic information.

The order of the objectives presented above should be reversed since, according to Fig. 3.3 illustrating the paradigm of vibroacoustical diagnostics, we have first the signal and then the vibroacoustic symptom. Nevertheless, the problem of the discrimination of vibroacoustical signals of wear is only at the beginning of its development; hence, we should start from the traditional objectives of a diagnostic experiment. The remaining two objectives will be discussed in Sections 6.4 and 6.5.

Such traditional objectives of an experiment in diagnostics are the answers to the questions: What?, Where?, and: How to measure?, in order to obtain information about the condition of the object and, more precisely expressed to obtain the constitutive relation of the object's diagnostics: 'symptom→ condition' relation. A view on the possible methods of obtaining such relation is presented in Fig. 6.1 from Żółtowski [108].

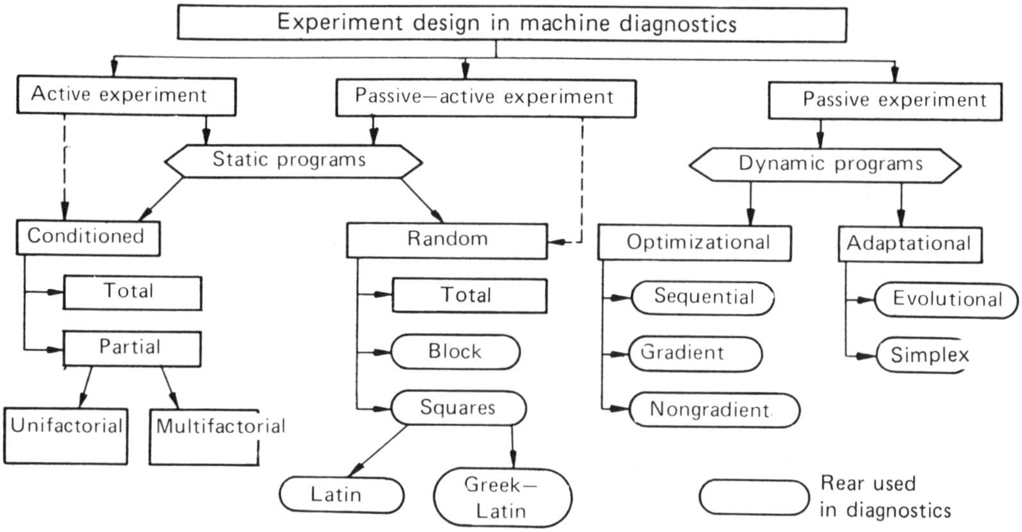

Fig. 6.1 — Systematics of experiments possible in machine diagnostics [108]

An active experiment was applied in the diagnostic investigation of mass-produced internal combustion engines, and it was carried out by the scientific workers of the Military Technical Academy [109, 110] and generalized by Krzyworzeka [111] and Adamczyk [112] in the example of piston compressors. These studies indicate a high utility of this type of experiment in the determination of the 'condition → signal' relation in cases where it is possible to carry out an experiment in laboratory conditions with a controlled change of condition and control variables. However, for mechanical objects, a controlled change of object's technical condition requires at least a disassembly of the cooperating assemblies, kinematic pairs, which next to organizational problems gives a different unit of the machine after reassembly. However, when we have a great number of such objects for investigation and a good laboratory, we can exclude the disturbances of the condition after disassembly by choosing an adequate form of an active experiment[112]. Not all possibilities have actually been used. For example, there are not trials with the application of canonical correlation [113] between the variables of the condition and vibroacoustical symptoms.

However, such form of diagnostic experiment is possible for multidimensional objects, and particularly for unitary objects working in process lines. Here, the only alternatives are the *active–passive* and the *passive experiments*. In the first case we can control the condition before starting the object and after stopping it (if there was no failure), and in the second we observe only the vibroacoustical effects of the natural (during use) wear of the object. Here, nothing else can be done but to make an hypothesis about the wear of the object and the vibroacoustical signal generation, and on this basis we can optimize the information gained from the experiment.

For a passive–active experiment we can still determine the condition → symptom relation having a partial knowledge of one-point or two-point condition parameters. On the other hand, in a passive experiment our constitutive equation of the object's diagnostics is described only in the field of utilization time by the life curve (or curves) $S(\theta)$. An exemplary illustration of the result of a passive–active experiment can be found in the doctoral dissertation by Żółtowski [114] as well as of a passive experiment [108]. The latter work presents an interesting idea of an adaptive construction of a diagnostic model of an object and the selection of measurement points. Since, in a passive experiment, we have no numerical information about the condition of the object, it has been proposed here to regard the main components of the observation matrix (Section 6.3) as a model, and for these components the next limit values have been determined as those necessary for a two-class determination of the good/faulty conditions. The idea is recommended for further application, particularly for unitary objects such as turbine sets. It requires some amount of computer processing of signals and information, but with the present possibilities it can be carried out by any mini- or microcomputer, or similar, to lead to the important contribution to the following.

(1) *Computer aided design (CAD).*
(2) *Computer aided manufacturing (CAM).*
(3) *Computer aided machine health monitoring (CAMHM)*
(4) *Computer aided maintenance and/or management (CAM).*
(5) *Computer aided engineering (CAE)*

6.3 DISCRIMINATION OF VIBROACOUSTICAL CHARACTERISTICS OF CONDITION AND CLASSIFICATION IN A PASSIVE EXPERIMENT

The possibilities of investigations according to the passive experiment in diagnostics are very limited; hence, greater attention should be devoted to the creation of diagnostic models in a passive experiment. And the creation of a model requires in each case the knowledge of vibroacoustical symptoms of wear.

The problem of identifying essential (distinctive) features from a given experiment was described for the first time in the 1930s in an agricultural experiment and in biology. The most important achievements include the works by Fisher and Anderson [113] which are the bases of the present knowledge about the discrimination of features. The achievements in this field were summarized in the monographs by Sobczak and Malina [115] and by Krzyśko [116]. In 1979, a discrimination algorithm [117] called BEDIND was developed, which is a passive experiment of diagnostic information detection whose computer version renders beneficial services in the Author's research group. Regardless of the fact that it has been presented by the Author [7, 117], it is worth quoting its outlines here and the possibilities and effects of its application.

6.3.1 BEDIND—algorithm of discrimination and classification

The passive diagnostic experiment can be of two types. The first concerns the occasion when in a selected point (points) of an object a certain number $i = 1, \ldots, s$ of vibro acoustical symptoms $S_i(\theta)$ is measured in selected moments of the lifetime of the object θ_j, $j = 1, \ldots, n$. In this way we obtain an observation matrix $Q = \{S_{ij}\}$ whose columns create the values of the particular symptoms, and the rows are arranged in the sequential order of time realization θ_j. Therefore, we can say that the realization of symptom $S_i(\theta)$ gives a vector which is the ith column of observation matrix Q. On the other hand, row Q is the realization of observation vector $\{S\} = \{S_1, \ldots, S_s\}$.

The second randomized version of the passive experiment consists in the simultaneous (in θ scale) observation of n units of objects of the same type with the same assumption that the following distribution of their lifetimes θ_j is uniform. In this way we obtain 'statistically the same' covering of the lifetime axis θ as in the previous case. For each unit of the object $j = 1, \ldots, s$ we measure $i = 1, \ldots, s$ of symptoms obtaining again the observation matrix $Q = \{S_{ij}\}$ whose columns are *symptoms, and the lines are the realization of symptoms for the successive units of machines* j.

Let us further assume that each symptom or its realization vector has the mean value $\bar{S}_i$ equal to zero, i.e. it is centered as below:

$$i\; E_j\left[S_{ij} - \bar{S}_i\right] = 0 \qquad \bar{S}_i = \frac{1}{n}\sum_{j=1}^{n} S_{ij} = E_j\left[S_{ij}\right] \tag{6.1}$$

Let us also standardize each vector to its standard deviation σ_{si} since wee will compare symptoms with physically different values and units. So, we have:

$$\sigma_{si} = \sqrt{E_j\left[S_{ij} - S_i\right]^2}, \qquad S_{ij}^{c} = \frac{S_{ij} - S_i}{\sigma_{si}}, \tag{6.2}$$

$$i = 1, \ldots, s$$
$$j = 1, \ldots, n$$

where S_{ij}^{c} is the element of the corrected in this way, i.e. standardized and centered observation matrix $Q_c = \{S_{ij}^{c}\}$.

The matrix of correlation (or covariance if without standardization) of the experiment can be obtained directly by multiplying the observation matrix by its transposition T since:

$$\sum_{c} = Q_c Q_c^{T}, \quad Q_c = \left\{S_{ij}^{c}\right\} \tag{6.3}$$

Eigenvalues λ_i of this correlation matrix can be found by solving the determinantal equation:

$$\det\left[\sum_{c} - \lambda_1 I\right] \equiv |\sum_{c} - \lambda_1 I| = 0 \tag{6.4}$$

where I is the diagonal unit matrix, and λ_i, $i = 1, \ldots, s$ are the sought-for solutions.
The eigenvalues arranged in a decreasing order:

$$\lambda_1 > \lambda_2 > \lambda_3 \ldots > \lambda_s \tag{6.5}$$

give information about the range of variability (ZZ) of each independent feature contained in our observation matrix Q_c.

The information content of each feature in our experiment can be determined in percentages in relation to the whole range of variability of the experiment

$$\sum_{i=1}^{s} \lambda_i = ZZE .$$

So we will have the participation of the particular independent features of our passive diagnostic experiment of Equation (6.5):

$$\lambda_1; \; \underline{ZZ}_i = \frac{100\lambda_i}{\sum_1^s \lambda_i} \%, \quad i = 1, \ldots, s$$

Since, in each case our experiment has a statistical character, the latter eigenvalues which are zeroes (literally or on the preset level of significance) [118, Section 4.4] must be rejected as statistically insignificant. In this way we determine at the same time a *statistically significant number of observed diagnostic features*, or independent failures which have been detected by our passive diagnostic experiment.

Now, there is the question in which way the independent features (failures) can be described with the help of the components of observation vector $\{\mathbf{S}\} = \{\mathbf{S}_1, \ldots, \mathbf{S}_s\}^T$, i.e. with the help of symptoms measured in the experiment. And so, for each eigenvalue λ_i there is a corresponding main component of observation matrix Y_i which is an element of a new observation vector $\{\mathbf{Y}_i\}$:

$$\{\mathrm{Y}_i\} = \{\mathrm{M}_i\}^T \{\mathrm{S}_i\} = m_{i_1} S_1 + m_{i_2} S_2 + \ldots m_{ir} S_r, \tag{6.6}$$
$$i = 1, \ldots, r < s$$

under the condition of normalization:

$$\hat{i} \{\mathrm{M}_i\}^T \{\mathrm{M}_i\} = 1$$

which carries information only about the i th independent damaging mechanism of the mechanical object. Here the vectors $\{\mathbf{M}_i\}$ are the eigenvectors of the correlation matrix as the result of the solution of the eigen problem for each already known λ_i:

$$\left[\Sigma_c - \lambda_i I\right]\{\mathrm{M}_i\} = 0, \quad i = 1, \ldots, r < s \tag{6.7}$$

where 'r' is the number of statistically significant elements λ_i . As we saw in Equation (6.6), the elements of the i th main component, due to the normalization to unity, are weights which help to create information about the i th way of damaging from the components of observation vector $\mathbf{S}_i$, $i = 1, \ldots, s$ measured in the experiment.

Therefore, if $V_t\, m_t \approx 0$, we can say that the component S_t of observation matrix Q

can be omitted and be omitted and on this basis the BEDIND program operates, evaluating the loss of information by calculating the range of the variability of experiment, the determinant of correlation matrix $\sum_c$ and excessive information = redundancy R of the experiment:

$$ZZE = \sum_{i=1}^{s} \lambda_i; \quad |\Sigma_c| = \prod_{i=1}^{s} \lambda_i; \quad R = \lg_2 \sqrt{\frac{\prod_{i=1}^{s} \sigma_{si}^2}{\prod_{i=1}^{s} \lambda_i^2}} \text{ bits} \tag{6.8}$$

Generally, in further investigation of the passive experiments we can cancel or ignore such component S_t , whose rejection does not change significantly the range of experiment variability and minimizes redundancy R, i.e.:

$$ZZE\,(S_1, \ldots, S_t \ldots) = \max \quad \text{and} \quad R\,(S_1, \ldots S_t, \ldots) = \min \tag{6.9}$$

$$t = 1, \ldots, s \qquad\qquad i = 1, \ldots, s$$

then, with the successive cancelling of the values of determinant $|\Sigma_c|$ (being the measure of the amount of independent information) see it grow very quickly close to unity.

The interruption of the process of experiment size reduction by successive cancelling of less important components S_t of the original observation vector must take place externally by the following means.

(1) Presetting the minimal number of components of the original observation vector, e.g. cancellation of up to three components.
(2) Presetting the maximum loss of the range of experiment variability due to cancelling.
(3) Presetting redundancy which can be permitted in the given experiment, e.g. $R = 2$ bits.

At each stage of the reduction of experiment size the BEDIND program calculates the values of features Y_i for each unit of the object $j = 1, \ldots, n$ (or for each observed moment of the lifetime θ_j). In this way we obtain numerical values of the detected features (failures) Y_i for each object, i.e. Y_{ij} :

$$Y_{ij} = \sum_{p=1}^{s} m_{ip} S_{pj}, \qquad i = 1, \ldots, r < s, \quad j = 1, \ldots, n \tag{6.10}$$

which is the basis for the classification in relation to the value of the detected features. Here, m_{ip} denotes the weight factors or eigenvectors of the i th feature, while S_{pj} constitutes the successive components of the original observation vector $\mathbf{p} = 1, \ldots, s$ in its j th realization. Further, treating the values Y_{ij} as the points which are the ends of vectors in a new r-dimensional orthogonal area of observation, we can calculate the lengths of observation vectors for each object:

$$|Y_j| = \sqrt{\sum_{i=1}^{n} Y_{lj}^2}, \quad j = 1, \ldots, n \tag{6.11}$$

Such a measure of the type of observation vector length is also the basis for another type of classification. In a case where the object of experiment is the measure of vibration velocity in different points and directions on the machine body, the length of observation vector will present the measurement of vibration energy.

Finally, the ends of vectors representing the objects give a resultant centre of experiment with coordinates:

$$Y_i = \frac{1}{n} \sum_{j=1}^{n} Y_{ij}, \quad i = 1, \ldots, r < s \tag{6.12}$$

The distance of each object $j = 1, \ldots, n$ can be the next basis for classification. It is worth noting here that the two latter ways of classification take into consideration all features, giving the bases for joint classification of objects. On the other hand, the classification of objects according to Equation (6.10) will be different for each failure (failures), $i = 1, \ldots, r < s$. At the same time, it is obvious that these classifications have no sense if we observe one object for the successive moments of time θ_j. They can be helpful only for the control of the monotonicity of life curves, since in this case the detection of the feature permits its further plotting as the partial life curve of failure.

Finally, it is worth considering the wider theoretical bases of the BEDIND algorithm. Its basic idea was taken from the monograph by Anderson [113. Chapter 11] about the main components of the observed population of objects. It is also similar to the discrimination methods described in other monographs [115, 116, 118]. According to the Author, it is most similar to the Karhunen-Loeve method [62, Section 5.2.3] of decomposition into an orthogonal series or sequence of vectors. In both cases (BEDIND and Karhunen-Loeve decomposition) the basis of departure for the development is the correlation matrix (or the correlation function for random processes), its eigen values and vectors (functions). As is usually the case, these associations have come significantly later than the elaboration of the BEDIND algorithm.

The BEDIND program with the principle of operation described above has been used in the Dynamics and Machine Vibration Group since 1979, being continually improved with the acquired experience. For at least a partial presentation of its possibilities we present below two sets of results.

In the Tarnów Nitrogen Plant four compressor units operate coupled to a recuperation turbine driven by an electric motor with an increasing gear shown in Fig. 6.2(a).

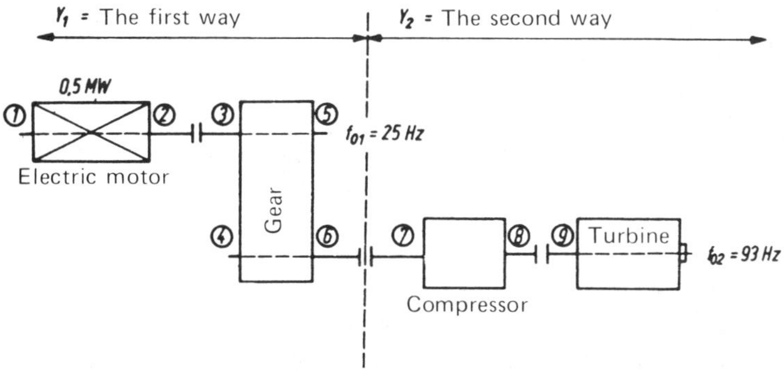

Fig. 6.2(a) — Kinematic diagram of a compressor unit with measurement points ①–⑨ of horizontal displacements of vibrations and with the indication of the separation of the mechanism of failures Y_1, Y_2 as the result of calculations according to the BEDIND program

The amplitudes of horizontal vibration displacements of bearings in points ①–⑨ were measured in irregular time intervals of the order of months. So it was not a temporal representation of life curves in points ①–⑨, but it was assumed that each measurement (altogether forty measurements were carried out) can be treated as the observation of another object (sufficiently changed by wear in relation to the previous measurement in this place). After processing the results * two significant components were obtained which are described below with a redundancy $R = 5.9$ bits:

$$\lambda_1 = 4.73; \quad ZZ_1 = 52\%; \quad Y_1 = 0.40\, X_{①} + 0.36\, X_{②} + 0.42\, X_{③} + 0.42\, X_{④} + 0.42\, X_{⑤} + 0.40\, X_{⑥}$$

$$\lambda_2 = 2.08; \quad ZZ_2 = 23\%; \quad Y_2 = 0.62\, X_{⑦} + 0.56\, X_{⑧} + 0.51\, X_{⑨}$$

So we see that up to the coupling between points ⑥ and ⑦ there dominates the first mechanism of damaging the unit Y_1, while after the second mechanism, Y_2 is dominating.

Looking at the classification of objects made in BEDIND program (Fig. 6.2(b)) we see that the greatest energy of vibrations is emitted by object number 8, and the smallest object 32 (we have already explained that the length of observation vector can be treated in this case as the measure of vibration energy).

The distance from the centre of our experiment gives another classification, since it is a differentiating measure form the *average object*. We see that objects that are in the middle of the previous classification (e.g. numbers 7 and 35) are now at the edge of this differential classification, while those that were previously the worst ones (8 and 21) remained in the same maximal distance form the middle of the experiment.

The main components Y_1 and Y_2 representing an independent mechanism of damaging the unit give quite different classifications. The first presents the technical condition of the unit on the drive line ①–⑥, the second presents the condition on the loading line ⑦–⑨ as in Fig. 6.2(a). It is possible now to calculate the limit value of each classification according to Section 4.3.3; this would facilitate at once the two-condition classification of the evaluated objects. Using the BEDIND program, we have a whole range of valuable indirect results about the intercorrelation of the components of the observation vector and their dynamics of change. It gives at once indications about what measurement magnitudes can be neglected without any essential loss of information. In the example we can preserve the components of the original observation vectors $X_{①}$, $X_{⑥}$, $X_{⑧}$ and $X_{⑨}$ giving a redundancy $R = 0.53$ bit without any significant change in the condition classification.

The successive example of the application of the BEDIND algorithm originates from the diagnostics of BAB-106 fans used in power engineering mentioned already (Fig. 5.3). In the Author's research group it has been frequently debated wether the use of a set of amplitudes of successive harmonics for the recognition of the condition would permit us to find a greater number of independent failures than is possible by the measurement of one vibrational value, e.g. the acceleration of horizontal vibrations of the bearing. It was still more encouraging since the acceleration amplitudes of the successive harmonics of rotation frequency f_0 up to $n = 12$ were characterized by statistically important changes.

* The author is grateful to F. Sordyl, PhD and R. Nowicki, PhD for making these results availbale.

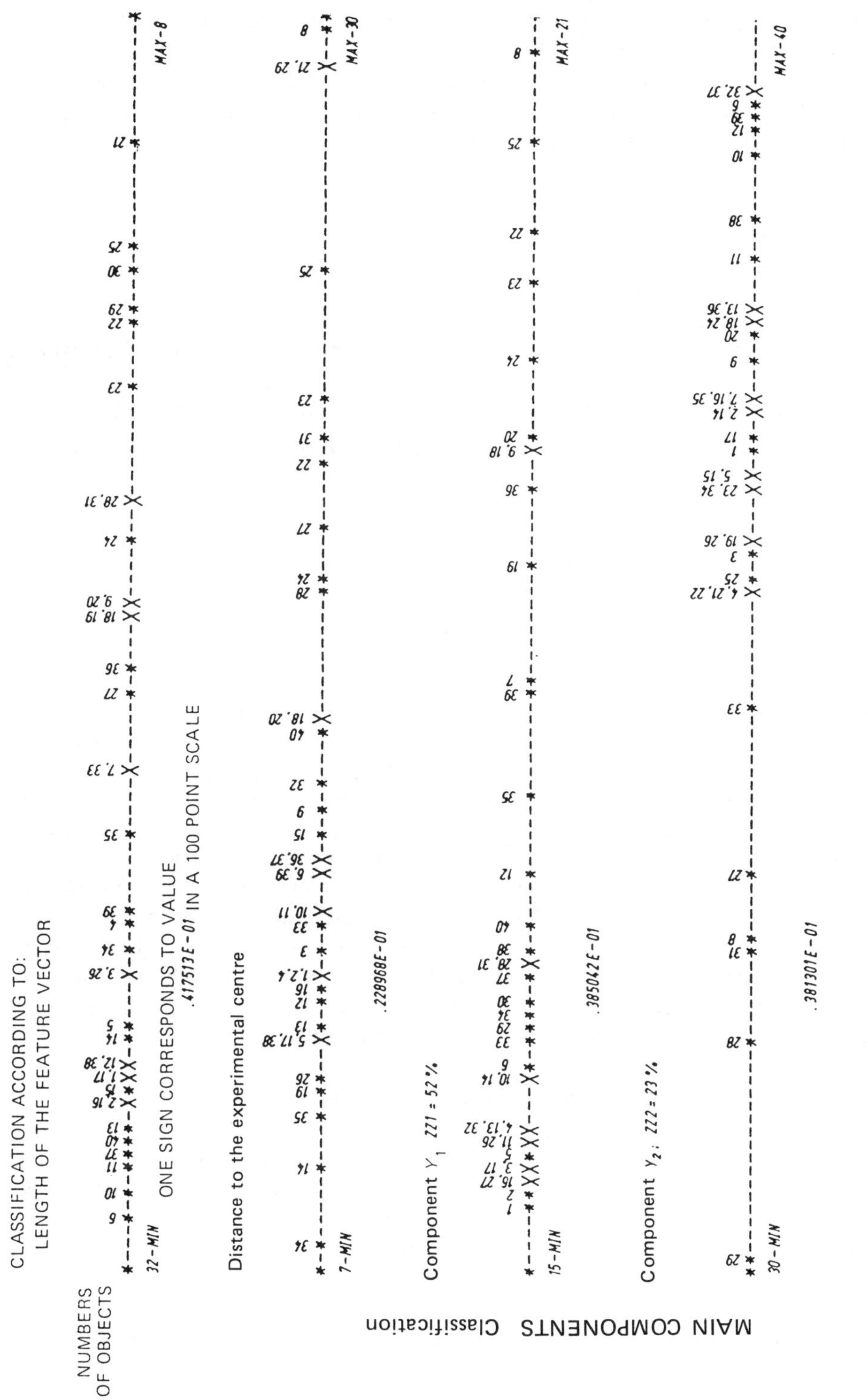

Fig. 6.2(b) — Classification of the condition of forty units of a turbocompressor according to the BEDIND algorithm

For the discrimination analysis the successive harmonics $n = 1, \ldots, 12$ of bearing vibration accelerations were collected: A = internal, and B = external according to Fig. 5.3 in the horizontal - H, and vertical = V directions for forty-one fans.

In each of the four directions of vibrations: AH, AV, BH, BV two or three statistically significant features were detected by BEDIND, i.e. the following mechanisms of fan wear with the following relative ranges of variability of the data:

Point AF;	$ZZ_1 = 36\%$,	$ZZ_2 = 15\%$,	$ZZ_3 = 12\%$
Point AV;	$ZZ_1 = 61\%$,	$ZZ_2 = 15\%$,	
Point BH;	$ZZ_1 = 93\%$,	$ZZ_2 = 6\%$,	
Point BV;	$ZZ_1 = 45\%$,	$ZZ_2 = 13\%$,	

As we see from the above, in each case, in spite of twelve harmonics, we can distinguish with certainty only two mechanisms of damage. The measurement of horizontal vibrations of the internal bearings AH of the fan gives the greatest information diversity, and the main components detected in this point are created in the following way:

$$Y_1 = 0.32\, a_1 + 0.31\, a_2 + 0.31\, a_3 + 0.26\, a_4 + 0.31\, a_5 + 0.27\, a_6 + 0.25\, a_7 + 0.11\, a_8 + 0.37\, a_9 + 0.04\, a_{10} + 0.28\, a_{11} + 0.4\, a_{12} \rightarrow \text{for } ZZ_1 = 36\%$$

$$Y_2 = -0.28 a_1 - 0.36 a_2 - 0.44 a_3 - 0.25 a_5 - 0.24\, a_9 - 0.42\, a_{10} - 0.46\, a_{11} + 0.25\, a_{12} \rightarrow \text{for } ZZ_2 = 15\%$$

i.e. generally:

$$Y_1 = \sum_{i=1}^{12} m_i\, a(i f_0); \quad Y_2 = -\sum_{i=1}^{6} m_i\, a(i f_0) + \sum_{i=7}^{12} m_i\, a(i f_0)$$

From the method of information creation about these failures cone can guess that the first feature indicates a total mechanical unbalance of fan $a\ (f_0)$, and aerodynamic unbalance $a\ (i \cdot f_0)$, $i = 2, \ldots, 12$. The second feature is the misalignment of the unit created as the weighted differences of the lower and higher amplitudes of harmonics. This hypothesis is also confirmed by a higher range of variability of this feature on the internal bearing (15%) than on the external one (6 and 13%). Hence, the internal bearing in the horizontal direction was accepted as the point of constant monitoring. Concluding the discussion on the BEDIND discrimination and classification algorithm it may be worth becoming acquainted with its many other applications given by the Author [7] and Żółtowski [108].

6.3.2 Discrimination of features by the orthogonalization method of observation vector components

The components of observation vector $S_i\ (\theta)$, $i = 1, \ldots, s$ are the amplitudes of physical magnitudes (e.g. of acceleration, velocity and vibration displacement) spectral components of these magnitudes, or other symptoms and discriminants of vibroacoustical processes (Section 6.3). In the application of the BEDIND procedure there develops the investigated features (failures) as the major components of the observation matrix. However, is it possible that some of the features can be measured directly in the experiment? It is essential, particularly in standard measurements where we are forced to use prescribed vibration values. Furthermore, the measuring traditions and the predispositions of the persons carrying out the investigations are sometimes not to be forgotten. Appreciating this problem, we have attempted to introduce into the BEDIND procedure some 'priority' symptoms, i.e. such symptoms which in spite of the indications of the procedure to discard the unnecessary measurement symptoms must remain until the end of the *algorithm*. Mathematically it is equivalent to imposing additional ties on the system, since the 'priority' interferes in the procedure of discarding. It is certainly a source of large errors, not easy to estimate. However, there exist procedures of orthogonalization of a known vector set (function set) in which the starting point can be chosen optionally. For example, it can be the amplitude of acceleration of vibrations recommended by the standard for measurement in the observed machine group while the remaining features orthogonal to the initial feature (here the acceleration) are chosen on the basis of a special Gram-Schmidt's algorithm [119, Section 3.3; 120, Section 14.7.4].

Let the realization of the measurement of each symptom 'i' always give us vector $\{\mathbf{S}_i\} = m\ \{\mathbf{S}_{ii}, \ldots, \mathbf{S}_{in}\}^T$, $i = 1, \ldots, s$. Let us assume further that we have placed the leading symptom as the first $i = 1$, e.g. the standard vibration acceleration. Then the quest for the orthogonal components of the new observation vector $\mathbf{W}_i$ has the following course:

$$\{\mathrm{W}_i\} = \{\mathrm{S}_i\}$$

$$\{\mathrm{W}_{i+1}\} = \{\mathrm{S}_{i+1}\} = \sum_{k=1}^{i} \frac{\{\mathrm{W}_k\}^T \{\mathrm{S}_{i+1}\}}{\{\mathrm{W}_k\}^T \{\mathrm{W}_k\}} \{\mathrm{W}_k\}, \quad i = 1, 2, \ldots \tag{6.13}$$

Cameron [119] noticed that the process of such orthogonalization can be accelerated significantly when vectors $\{\mathbf{W}_i\}$ zeroing on the preset level of significance are carried to the end of the initial set $\{\mathbf{S}_i\}$. In such cases the procedure of Equation (6.13) is carried out to that place in which all remaining vectors to be transformed prove to be zero ones.

The relation defining the new components of observation vector for the centred vectors $\{\mathbf{S}_i\}$, i.e. $\hat{i}\, E_j\{\mathbf{S}_i\} = 0$, has a first component of the sum which is nothing else but the correlation coefficient between the k the component of the newly created vector and the component of the original observation vector $\{\mathbf{S}_{i+1}\}$treated on the successive 'i + 1'th step. Therefore, if the treated vector $\{\mathbf{S}_{i+1}\}$ is orthogonal to the already calculated components of the new vector, i.e. if $\{\mathbf{W}_k\}^T \{\mathbf{S}_{i+1}\} = 0$, $k = 1, \ldots, i$ then $\{\mathbf{W}_{i+1}\} = \{\mathbf{S}_{i+1}\}$ and at this step the orthogonalization algorithm terminates.

So far, there are no attempts to utilize this simple algorithm for the discrimination of the features of the system of random variables created as a result of the observation of a group of objects oriented to a definite measurement magnitude. There are only some trials to utilize the orthogonalization algorithm to the handling of random processes [121] which will be discussed in Section 6.5.

6.4 RECOGNITION OF CONDITION

In Chapter 4 we discussed the most important diagnostic models of mechanical objects giving the assignment symptom → condition. In the covert model of the image type we discussed the geometrical method of recognizing the condition by the construction of the construction of the decision hypersphere. Next to this recognition method there are also the minimal-distance methods, and probabilistic and linguistic methods [84]. Therefore, it seems worth presenting a short discussion on the bases of the minimal-distance methods which find application in machine diagnostics (see, for example, the work of Cholewa (85]).

Imagine that we have two sets of symptom vectors creating matrices as follows.

(1) The training set—with a known condition assignment of k objects:

$$\left\{S^{u}_{ir}\right\}, \quad r = 1,\dots,k \quad u = \begin{cases} z \text{ in Good Condition} \\ n \text{ in Faulty Condition} \end{cases}$$

(2) A set of n objects for the recognition of their condition:

$$\left\{S_{ij}\right\}, \quad j = 1,\dots,n \tag{6.14}$$

where $i = 1, \dots, s$ are the components of the observation vector of each object from the training set $r = 1, \dots, k$ and for the set $j = 1, \dots, n$ being recognized.

Now, when we want to recognize the condition of the j th object, we calculate its distance d_{rj} from each object of the training set. On the basis of the knowledge of these distances we can make a judgement about the similarity of the recognized object to its closest neighbour $\left(\min\limits_{r} d_{jr}\right)$ or to another condition standard.

There exist many definitions of distances (but see, particularly, that by Cholewa [85]) and one of the more general definitions—that of Minkowski—has already been utilized here in the definition of vibroacoustical signal measures (Section 3.5). For our purposes it can be written in the following way:

$$d^{\mu}_{rj} = \left[\sum_{i=1}^{s} \left|S_{ij} - S^{u}_{ir}\right|^{\mu}\right]^{1/\mu} \qquad u = z, n \tag{6.15}$$

at the same time, owing to the different names of the symptoms S_i and differential scales, it is beneficial to standardize the features by dividing them by the standard deviation in the analysed set:

$$\sigma_i = \sqrt{\frac{1}{n}\sum_{j=1}^{n}\left(S_{ij} - \bar{S}_i\right)^2}, \quad \sigma^{u}_{i} = \sqrt{\frac{1}{k}\sum_{r=1}^{k}\left(S^{u}_{ir} - \bar{S}^{\bar{u}}_{i}\right)^2} \tag{6.16}$$

where $\bar{S}_i$, $\bar{S}^{u}_{i}$ are the mean values of symptoms:

$$\bar{S}_i = \frac{1}{n}\sum_{j=1}^{n} S_{ij}, \qquad \bar{S}^{u}_{i} = \frac{1}{k}\sum_{r=1}^{k} S^{u}_{ir}$$

The choice of the exponent μ as dictated by our needs; if we want to consider all features equally, then μ - 1; if we want to take into consideration the greater differences in the features, then we should accept $\mu > 1$, and for $\mu \to \infty$ we obtain

$$d^{u}_{rj} = \max_{i}\left|S_{ij} - S^{u}_{ir}\right|.$$

The distance d_{rj} is a measure of the closeness of objects and this measure can already be the basis for the recognition of the condition. Assuming, for example, that the j th object has the same condition as its closest neighbour, we can accept the following decision rule $D\ (s_j)$:

$$D(S_j) = \begin{cases} z, \text{ if } \min\limits_r d^z_{rj} < \min\limits_r d^n_{rj} \\ n, \text{ if } \min\limits_r d^n_{rj} < \min\limits_r d^z_{rj} \end{cases} \tag{6.17}$$

and accept decisions about the good condition of the j th object if its closest neighbour is in good condition and vice versa.

The decision rule is simple, but it requires, each time, the calculation of the distances from all objects of the training set S^u_{ir}, $r = 1, \ldots, k$, $u = z, n$. There exist many alternatives aiming at the decrease of the number of calculations with the preservation or even with the correctness of the previous classification quality.

However, we will be satisfied with the presentation of the general idea of the minimal distance classification and the method of the closest neighbour, and will not go into further details, nor discuss the probabilistic and linguistic methods. The interested readers are referred to the subject literature [84–86] and Tadeusiewicz [122].

In conclusion, it is worth mentioning the successful classification trial of internal combustion engines directly with the help of finite realizations of signals from transducers mounted on the engine as shown by Gersch, Brotherton and Braun [57]. Here, the method of the closest neighbour was used, and Kulback's divergence was applied as the measure (Equation (3.18)) whose merit is that it permits us to calculate at once the probability of an erroneous classification.

6.5 DISCRIMINATION OF FAILURE SIGNALS

The special techniques of discrimination and image recognition discussed above form an antidote for the high level of disturbances contained in the measurement symptoms $S_i\ (\theta)$. However, the problem of disturbance and noise can be approached in a different way, minimizing them after the reception of a signal and before the creation of a symptom. A simple treatment of this type, which was already mentioned, is the successfully applied band filtration, e.g. in the measurement of kurtosis of bearing vibrations (Section 3.5.3). However, in many cases, such rectangular characteristics of the band filter do not suffice and another filter is necessary which changes its transmittance in an adaptational way with (or without) some information about the nature of disturbances. The idea of such development—called *adaptive noise cancellers*—was been extensively presented by Widrow, *et al.* [123] in 1975. At present, we have undertaken successful diagnostic applications of adaptive noise cancellers and the miniaturization and increase of the calculation possibilities of microprocessors accelerates the development of this promising direction for diagnostics.

A significantly smaller calculation is required by the orthogonal filtration of signal (we were discussing it in connection with the discrimination of characteristics). On the other hand, the greatest number of calculations—but not in real time, as in the previous cases—are required by the diagnostic filtration being a combination of BEDIND and orthogonal filtration. The three discrimination techniques of failure signals are discussed below.

6.5.1 Adaptive noise cancelling

The need for an adaptive noise canceller of a vibroacoustical signal is best explained by referring to Equation (3.2) (which, in its final form becomes Equation (6.18)) of our signal generating model:

$$s(r, t, \theta) = \sum_{i=1}^{n} h(r_i, t, \theta) \otimes V_i(t, \theta) + c(r, t, \theta) + n(r, t, \theta) \tag{6.18}$$

Here, the original failure signals V_i (·) are generated in r_i places and in the convolution with impulse function of transmission path h (r_i, t, θ) (from the place of failure to the place of reception) coming to the reception point 'r', mixed further with the rated signal of machine c (r, t, θ) and the uncorrelated disturbances n (r, t, θ), If the subject of our investigation is the k th mechanism of damage and the signal connected with it V_k (t, θ), then we can write:

$$s(r, t, \theta) = h(r_k, t, \theta) \otimes V_k(t, \theta) + \sum_{i=1}^{n} {}^{\rho\, k}\, h(r_i, t, \theta) \otimes V_i(t, \theta) + c(r, t, \theta) + n(r, t, \theta) \tag{6.19}$$

$$= h(r_k, t, \theta) \otimes V_k(t, \theta) + m(r, t, \theta), \qquad \Sigma^{\rho\, k} = \text{no '}k\text{' component}$$

Using an *ideal* adaptive noise cancelling system we will obtain from the above a qualitative illustration of the principle of its operation:

$$\text{ANC}[s(r, t, \theta)] \cong V_k(t, \theta) \tag{6.20}$$

As we can see, an elimination of all noise and disturbance marked altogether as m (r, t, θ) has taken place here, and even a deconvolution of the impulse function of transmission path h (r, t, θ) occurred. The execution of the above idea is not so simple, but in many cases it is possible, provided we possess a programmed digital signals analyser. The idea of such operation in the application to diagnostics will be illustrated according to the works of Chaturvedi and Thomas [124, 125] and Figure 6.3 explains the flow diagram of the sampled signal.

It can be observed from Fig. 6.3 that there are two inputs; the first is that of the measurement signal with noise s (t) + n_0 (t), and the second that of the reference signal with noise n_1 (t) from which the output of the adaptive filter y is formed stepwise with the help of weights W_{ij} whose effect has the aim to minimize ε, i.e. the output of adaptive noise canceller. We assume here that the processes s (t), n_0 (t), n_1 (t) are stationary with zero mean value, and that the noises n_0 (t), n_1 (t) are mutually uncorrelated with the signal, i.e. E_t {$\mathbf{S} \cdot \mathbf{n}_0$} =0, E_t {$\mathbf{S} \cdot \mathbf{n}_1$} =0, E_t {$\mathbf{n}_1 \cdot \mathbf{n}_0$} =0.

The output of the adaptive noise canceller signal, as visible in Fig. 6.3, can be expresses by the formula:

$$\varepsilon = s + n_0 - y \tag{6.21}$$

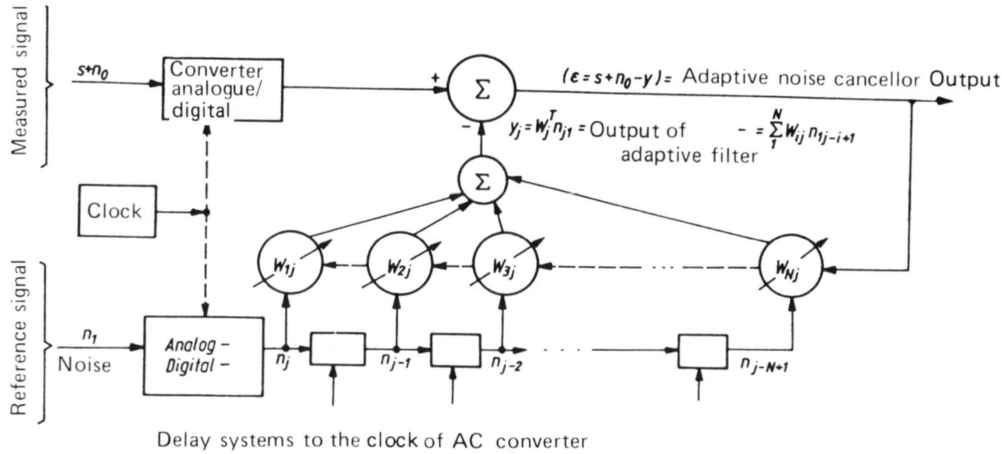

Fig. 6.3 — Exemplary configuration of an adaptive noise cancelling system [125]

$$\text{then}\quad \varepsilon^2 = s^2 + \left(n_0 - y\right)^2 + 2s\left(n - y\right)$$
$$= s^2 + \left(n_0 - W^T n_1\right)^2 + 2s\left(n_0 - W^T n_1\right) \tag{6.22}$$

Taking the expected value of both sides and using the assumption about the zero value of the signal mean $E_t\{S\} = 0$ we have:

$$E_t\left[\varepsilon^2\right] = E_t\left[s^2\right] + E_t\left[n_0 - W^T n_1\right]^2$$

Hence, we see that by minimizing the output ε by the selection of the weight vector **W**, we obtain an undisturbed signal s since:

$$\min_W E_t\left[\varepsilon^2\right] = E_t\left[s^2\right] + \min_W\left[n_0 - W^T n_1\right]^2 \tag{6.23}$$

then, as we see from the above the output of the adaptive filter $y = W^T n_i$ is the best estimation (in the sense of a mean square) of the disturbing noise n_0; thus, it is the best replica of these disturbances.

There are many methods of discovering the current values of weight vector $\mathbf{W}_j$ and we refer the reader to the literature [123, 125]. We will show only the effect of the action of the adaptive noise cancelling system in application to the diagnostics of the rolling element bearing. The measurement signal $s + n_0$ was collected from the bearing casing and the reference signal n_1 was taken from the base of the measurement stand consisting of an electric motor, a toothed gear and two rolling element bearings [125]. For the evaluation of the condition of rolling element bearings, the kurtosis β of each signal was measured, as in Fig. 6.4. The signal of failure appeared here at the output of the adaptive filter, giving the value of kurtosis β of 10.5, which is a proof of a significant failure of the bearing. Two conclusions are suggested here: (a) one cannot ensure the place of the signal appearance (ε, y) which is an unfavourable characteristic; (b) there follows a significant increase of the signal reaching a noise level of the order $30\text{dB}, \left(20\lg\dfrac{2.1\cdot 10^{-1}}{7.7\cdot 10^{-2}} = 31.3\text{dB}\right)$

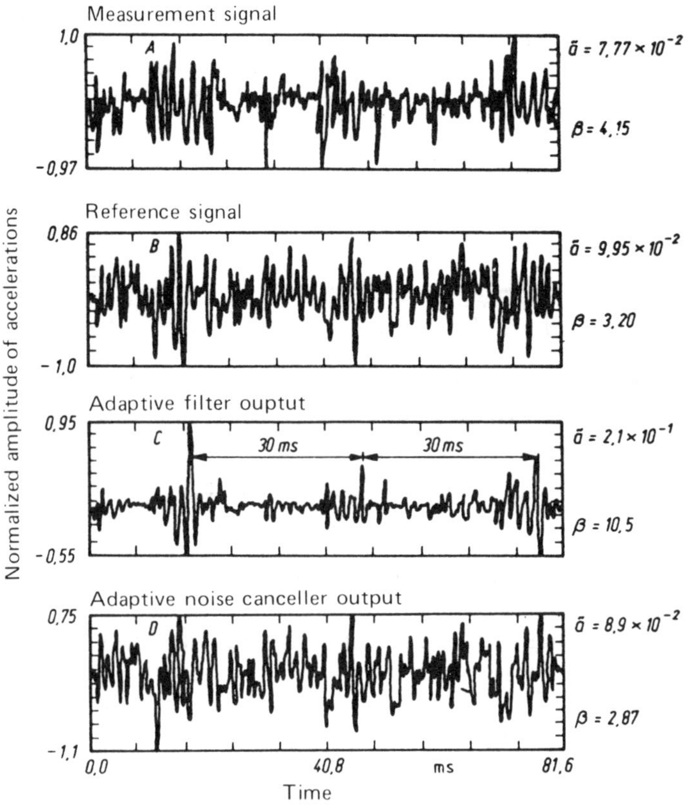

Fig. 6.4 — Illustration of the action of the adaptive noise canceller in a standardized scale of acceleration with an indication of the corresponding value of kurtosis β [125]

permitting without difficulty the evaluation of the condition of the bearing with the help of kurtosis measurement. So we see here that our considerations about the preliminary selection of signals (Section 3.4) were most justified.

6.5.2 Orthogonal filtration of signals

In some diagnostic cases of signal analysis we do not possess such complicated apparatus as that required for adaptive noise cancelling. However, at least a transformation of the received signal is advisable for the extraction of the presumed independent characteristics, i.e. the discrimination of failures. The orthogonalization of the components of the observation vector presented earlier in Section 6.3.2 permits a direct transfer of this method to the class of signals. Considering how to select the components of the observation vector, we can select for vibrations the successive derivatives of the vibration process: $x(t) = x^{(0)}$, $x(t) = x^{(1)} = v(t)$, $x(t) = x^{(2)} = a(t)$, and $x(t) = x^{(3)} = j(t)$, i.e. the displacement, velocity, acceleration and jerk of vibration or, conversely, the successive integrals of vibration jerk.

Using the results taken from the work of Kowalski [121], we can present two orthogonal systems of vibration signals with an ascending order of derivatives:

$$
\begin{aligned}
W_0(t) &= x^{(0)}(t) = x(t) \\
W_1(t) &= x^{(1)}(t) = v(t) \\
W_2(t) &= x^{(2)}(t) - \left(\frac{\tilde{v}}{\tilde{x}}\right)^2 \cdot x(t) = a(t) - 4\pi^2 f_x^2\, x(t) \\
W_3(t) &= x^{(3)}(t) - \left(\frac{\tilde{a}}{\tilde{v}}\right)^2 \cdot v(t) = j(t) - 4\pi^2 f_v^2\, v(t),
\end{aligned}
\tag{6.24}
$$

fx , fv = Rice frequencies

and with a decreasing order of vibration signal derivatives:

$$
\begin{aligned}
W_0(t) &= x^{(3)}(t) = j(t) \\
W_1(t) &= x^{(2)}(t) = a(t) \\
W_2(t) &= x^{(1)}(t) - \left(\frac{\tilde{a}}{\tilde{\gamma}}\right)^2 x^{(3)}(t) = v(t) - \frac{1}{4\pi^2 f_a^2}\,\gamma(t) \\
W_3(t) &= x^{(0)}(t) - \left(\frac{\tilde{v}}{\tilde{a}}\right)^2 v^{(2)}(t) = x(t) - \frac{1}{4\pi^2 f_v^2}\,a(t)
\end{aligned}
\tag{6.25}
$$

In introducing these relations we can take Equation (6.13) as the basis, and make some statements about the correlation coefficients of process derivatives [77, 121] i.e. that:

$$
\begin{aligned}
&E_t\left[x^{(i)} \cdot x^{(i-2k)}\right] = E_t\left[x^{(i-2k+1)}\right]^2 \\
&E_t\left[x^{(i+1)} \cdot x^{(i)}\right] = E_t\left[x^{(i+k+1)} \cdot x^{(i-k)}\right] = 0 \\
&i = 1,\ldots,n, \quad k = 1,\ldots, i-k \geq 0
\end{aligned}
\tag{6.26}
$$

and the known formula for Rice's frequency of acceleration $f_a = \dfrac{\tilde{\gamma}}{2\pi\,\tilde{a}}$, velocity $f_v = \dfrac{\tilde{a}}{2\pi\,\tilde{v}}$, and displacement $f_x = \dfrac{\tilde{v}}{2\pi\,\tilde{x}}$, where the tilde ~ indicates the root – mean – square value of the given magnitude

As stated elsewhere [121], the first system of signals can be used for the diagnostic observation of low-speed machines, while the second one can be used for the high-frequency phenomena (e.g. for the diagnostics of rolling element bearings) creating in each case the measures of signals described in Section 3.5. The above component signals of the observation vector in Equations (6.24) and (6.25) must be independent in two fields: (a) in signal description, (i.e. in dynamic time T); and (b) in the field of the characteristics of the observed objects, or the amplitude measures in Equation (6.24) and (6.25). For the demonstration of the above [121] the root-mean-square values of the components of Equation (6.24) were calculated from the original data (a, v, x) of twenty-eight rear axles of BRDM vehicles. The original correlation between the sets of measured symptoms was:

$$\rho(\tilde{a}, \tilde{x}) = 0.5, \qquad \rho(\tilde{a}, \tilde{v}) = 0.75$$

and after orthogonalization, in which according to Equation (6.24) only $a(t)$ was subject to change into $W_2(t)$, we have:

$$\rho(\tilde{W}_2, \tilde{x}) = -0.08, \qquad \rho(\tilde{W}_2, \tilde{v}) = -0.3$$

Therefore, the decorrelation of amplitude measures is significant and is worth being recommending for application in the field of symptom measurement.

6.5.3 Diagnostic filtration of signals

After the presentation of the favourable properties of the discrimination of features and disturbances in the field of symptoms, as well as the adaptive filtration of failure signals, a thought is suggested to connect both values in the field of symptoms and signals in a passive diagnostic experiment. The simplest diagnostic observation model of the object with a measurement of vibrations in one previously selected point 'r' can have the form as shown in Fig. 6.5.

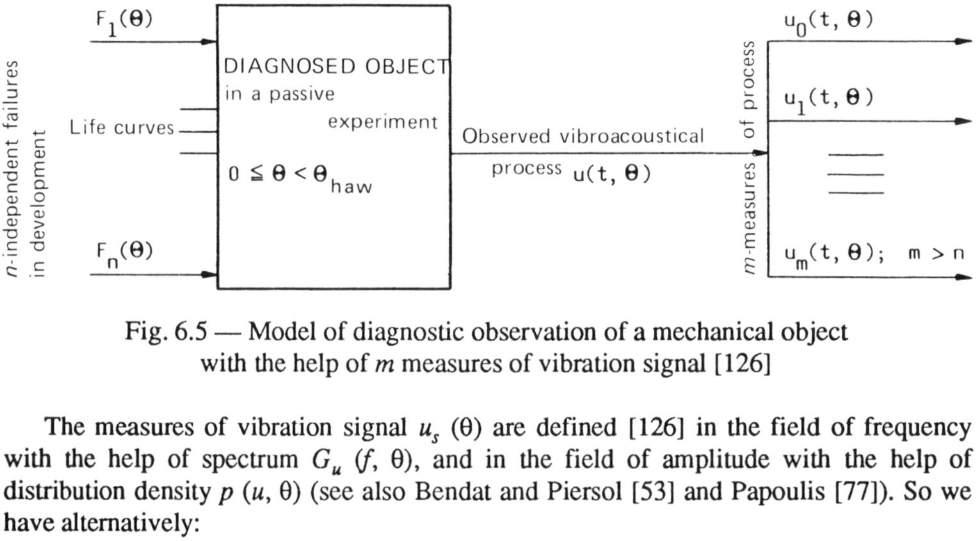

Fig. 6.5 — Model of diagnostic observation of a mechanical object with the help of m measures of vibration signal [126]

The measures of vibration signal $u_s(\theta)$ are defined [126] in the field of frequency with the help of spectrum $G_u(f, \theta)$, and in the field of amplitude with the help of distribution density $p(u, \theta)$ (see also Bendat and Piersol [53] and Papoulis [77]). So we have alternatively:

$$u_s(\theta) = \begin{cases} \int\limits_0 f^{2s} G_u(f,\theta)\, df = E_t\left[\overset{W}{U}{}^{(s)}(t,\theta)\right]^2 & \text{– spectral observation} \\ & s = 0,1,2,\ldots, m > n, \\ \int |u|^{s+1} p(u,\theta)\, du = E_t\left|\overset{A}{U}(t,\theta)\right|^{s+1} & \text{– amplitude observation} \end{cases} \qquad (6.27)$$

The successive realization of the composition of the observation vector of $u_s(\theta_j)$, $s = 1, \ldots, m$, $j = 1, \ldots, l > m$ will give us lines of two alternative matrices of observation $Q_c = \{\mathbf{U}_{sj}\}$ as the input of the BEDIND procedure (Section 6.3.1). Marking the observed failure characteristics (main components) in the spectral alternatives by

$$\overset{W}{D}{}^2_r(\theta)$$

according to Equations (6.6) and (6.27) we can write:

$$\overset{W}{D}{}^2_r(\theta) = \sum_{s=0}^{m} b_{rs} \overset{W}{U}_s(\theta) = \sum_{s=0}^{m} b_{rs} \int_0 f^2 G_u(f,\theta)\, d\theta, \quad r = 1, \dots, p_W < n \qquad (6.28)$$

Similarly, for the alternative of amplitude observation the main components of the experiment will be:

$$\overset{A}{D}{}^2_r(\theta) = \sum_{s=0}^{m} d_{rs} \overset{A}{U}_s(\theta) = E_t\left\{ \sum_{s=0}^{m} d_{rs} \left|u(t,\theta)\right|^{s+1} \right\}, \quad r = 1, \dots, p_A < n \qquad (6.29)$$

where P_W and P_A are numbers of significant features at the present level of significance.

We can guess from the above that in both variants of experiment, the spectral W and amplitude A ones, we distinguish the same features; however, it will be worse regarding their mutual range of variability and ordering.

Let us look creatively at the features found from the spectral experiment in Equation (6.28). Recalling the principles about the transition of signal with a known spectral density $G_u(f, \theta)$ through a linear system, we can treat the expression standing lower at the spectrum as the square of transmittance module $\phi(if)$ of the ϕ linear filter system since:

$$\overset{W}{D}{}^2_r(\theta) = \int_0 \left(\sum_{s=0}^{m} b_{rs} f^{2s} \right) G_u(f,\theta)\, df = \int_0 \left|\phi_r(if)\right|^2 G_u(f,\theta)\, df \qquad (6.30)$$

$$r = 1, \dots, p_W < n$$

The above expression is a constitutive relation of the r th filter oriented to the r th failure in the field of the spectrum; it is seen to be the averaged square of the signal of the r the feature $D_r(t, \theta)$ since:

$$\overset{W}{D}{}^2_r(\theta) \equiv E_t\{D_r(t,\theta)\}^2 = \int_0 \left|\phi_r(if)\right|^2 G_u(f,\theta)\, df \qquad (6.31)$$

which follows directly from the theorem of Parseval [53, Section 3.5]. The synthesis of filter with the preset transmittance is a problem of electronic circuity; therefore, we will not go further, knowing that in the result of a passive experiment and discrimination of spectral features we can obtain a disturbed signal of the r th mechanism of damage:

$$D_{rc}(t,\theta) = D_r(t,\theta) + N_r(t,\theta) \qquad (6.32)$$

where $N_r(t, \theta)$ is the disturbance of signal after discrimination filtration.

A further improvement of the relation signal to noise $[D_r(\theta)/N_r(\theta)]$ can be obtained by the momentarily forgotten amplitude alternative of the passive experiment according to Equation (6.29) used in series after spectral filtration.

For a qualitative illustration of a coupled spectral–amplitudinal filtration of vibroacoustical signals let us consider the analytical example of spectral observation of a mean square velocity and acceleration of object in the representative point$\{\overline{v^2(\theta)}; \overline{a^2(\theta)}\}$and amplitude observation of the first and second moment of vibration velocity $|v(\theta)| = \overline{v(\theta)}$,

$\overline{|v(\theta)|}^2 = \overline{v^2(\theta)}$, BEDIND applied to the field of spectrum gave the following main components:

$$W_{D_1}^2(\theta) = b_{10}\overline{v^2(\theta)} + b_{11}\overline{a^2(\theta)};\quad W_{D_2}^2(\theta) = b_{20}\overline{v^2(\theta)} + b_{21}\overline{a^2(\theta)},\quad b_{ij} > \theta$$

hence (6.31)

$$W_{D_1}^2(\theta) = \int_0 \left(b_{10} + b_{11}f^2\right) G_v(f,\theta)\, df,\quad W_{D_2}^2(\theta) = \int_0 \left(b_{20} - b_{21}f^2\right) G_v(f,\theta)\, df$$

The spectral and signal realization of both main components is presented in Fig. 6.6.

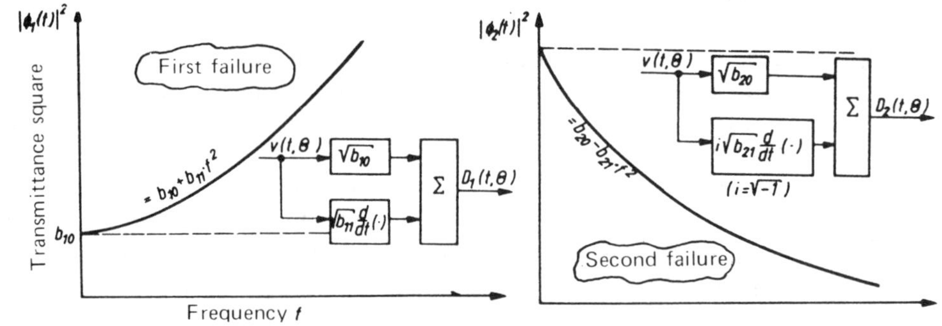

Fig. 6.5 — Square of transmittance and the simplest structure of filters diagnostically oriented [126]

Making now a synthesis of amplitude filter and connecting in a series we will obtain a coupled method of failure signal discrimination of our hypothetical machine, as in Fig. 6.7.

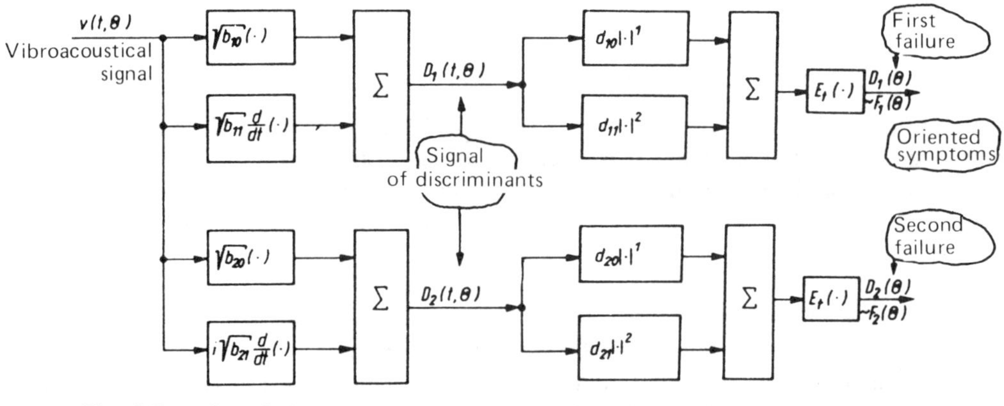

Fig. 6.7 — Coupled spectral–amplitude filtration of vibration velocity signal of a hypothetical machine for the detection of failure discriminants $D_1(\theta)$, $D_2(\theta)$ [126]

This method of coupled filtration can be called the diagnostic filtration. Theoretically, it is very elegant, but one link mentioned earlier is lacking, i.e. the assignment of failures in the fields of spectrum and amplitude. It can be done for example, by studying the similarity of life curves ${}^{W}D_{r}^{2}(\theta)$ and ${}^{A}D_{r}^{2}(\theta)$ for $r = 1,\ldots,$.

So far, the above conception has not been fully verified experimentally [126].

6.6 SUMMARY

In this chapter, we have traced the main direction in which the extension of knowledge and diagnostic possibilities take place. The majority of effort is concentrated around the passive experiment, and its good planning and execution permit the discrimination of features and the classification with the help of the BEDIND procedure of the orthogonalization of observation vector components, or the evaluation of the condition by one of the methods of image recognition.

An improvement in the signal to noise ratio in the field of symptoms, as mentioned above, is in many cases sufficient. However, a significantly better solution is the preliminary filtration of failure signal and then the measurement of the failure-oriented symptom. The most advanced method of such filtration is that adaptive noise cancelling and sometimes, also, orthogonal filtration, being instrumentally significantly simpler.

The diagnostic filtration which is based on the discrimination of failure signals and the serial measurement of failure-oriented moments of signals has not yet acquired any solid experimental basis for the evaluation of this method. Nevertheless, it seems that the presentation of the method of inference has been valuable, and the same applies to all methods presented in this chapter.

Appendices

The reader must have observed that this book has more the character of method than an instrumental approach. Hence, in the main text there is no information which in the light of the chosen line of argument would be superfluous. In order to give a complete picture of the problem, we present now some data and information which can have significant importance in plants applying the vibration diagnostics of machines. In the first place, we are going to present the important standards of vibration diagnostics which should be consulted when diagnostic problems are to be solved. Then, we will present in a condensed form the diagnostic chart containing information about machine behaviour facilitating the organization of the diagnostic process as well as the localization of the failure. Finally, we are going to present the criteria of balancing turbomachines, and a conversion nomogram of vibration amplitudes.

APPENDIX 1—IMPORTANT STANDARDS OF MACHINE VIBRATION DIAGNOSTICS

Section 3.5.2 (Fig. 3.24) gives the diagnostic standard of the International Organization for Standardization formulated in the categories of root-mean-square value of vibration velocity $\tilde{v}$ for four-machine groups. These are not the only standards, and furthermore, the root-mean-square amplitude of vibrations is not a diagnostically predestined value. For a complete understanding of the diagnostics of all types of machines and installations we need standards formulated in peak amplitudes of displacements for low-speed machined, in peak amplitudes of vibration velocity for general machines and installations, and in peak amplitudes of acceleration for toothed gears and rolling element bearings.

An American diagnostic company IRD Mechanalysis published their own standards in this respect and they deserve to be quoted.

Such a standard formulated in the categories of the peak value of displacement and vibration velocities measured on the machine bearings housing by a filter set to its rotary frequency f_0 is presented in Fig. A.1. However, the unfiltered limit readouts are very

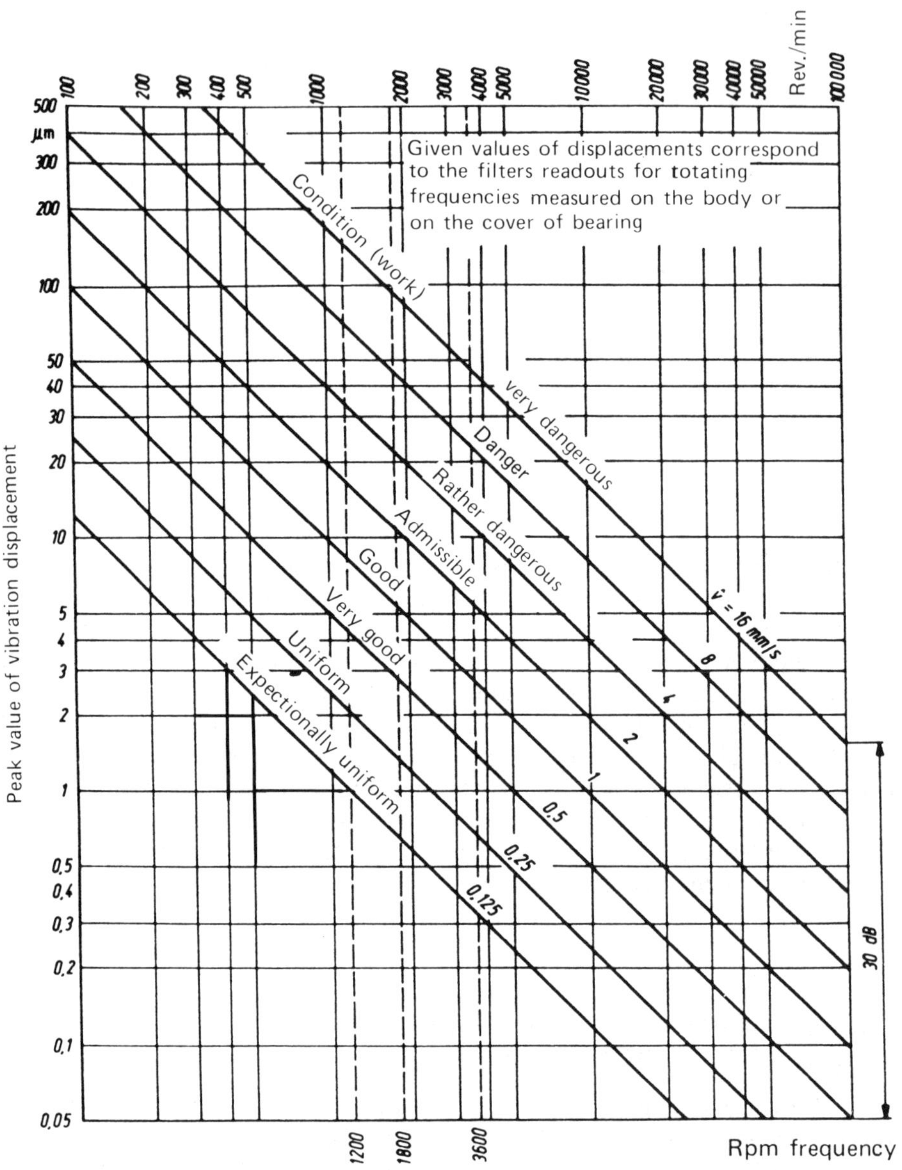

Fig. A.1 — Diagnostic evaluation of machine vibrations according to IRD Mechanalysis [67]

close to the values given in Fig. A.1 [65] which should be remembered.

Slightly different data are presented by the British company VCI Ltd (Fig. A.2). It is a mixed standard giving the peak values of vibration displacement and the root-mean-square values of vibration velocity. Comparing both standards, we can say that, in spite of this difference, they are very similar in their upper limits; IRD: $\hat{v}$ =16 mm/s, and VCI: $\tilde{v}$ = 15.2 mm/s, etc. Therefore, they could be used interrchangeably remembering that the VCI standard does not require any filtration in rotational frequency [11].

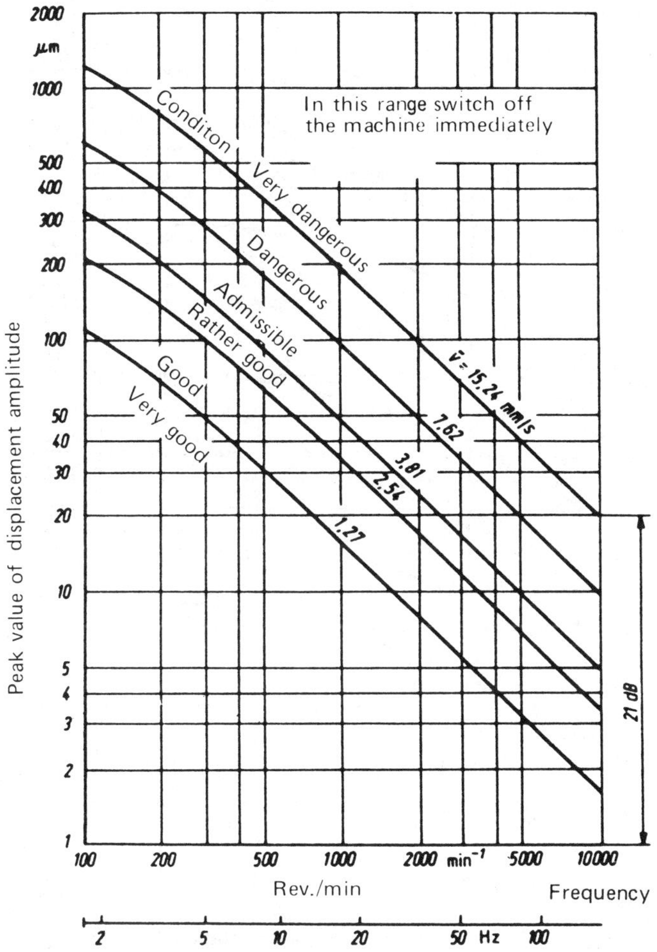

Fig. A.2 — Vibrational limit conditions of machime according to the determinations of the British company VCI Ltd [11]

For high-speed machines, rolling element bearings and gears it is beneficial to diagnose according to the readouts of accelerations. The standards used by the IRD are given in Fig. A.3. As we see, it permits us to use the peak values of velocity or acceleration in determining the vibration condition of a machine.

The successive standards presented in the categories of vibration velocity are shown in Fig. A.4. It is an extension of standard ISO 2372 for big and heavy machines, and as is usual for ISO, they are given in root-mean-square amplitudes of vibration velocities.

The most universal evaluation of machine vibrations is presented in the standard given by Blake [770] formulated for peak values of velocity, displacement and acceleration with the possibility of expansion of the amplitude scale with the help of service factor D, whose typical values are given under Fig. A.5.

Based on Blake's standard, in the Dynamics and Vibroacoustics of Machines Group Poznan *UT*, a diagnostic slide rule has been constructed for the evaluation of machine

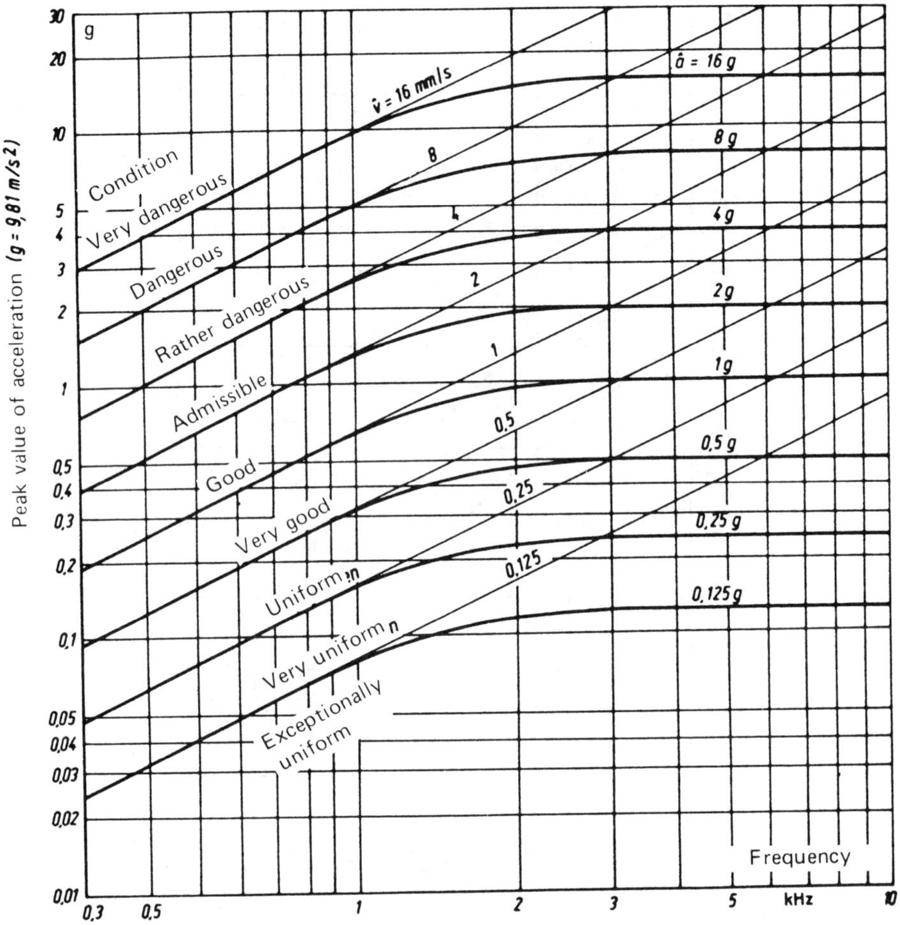

Fig. A.3 — Vibrational condition of machines in the categories of peak velocities and vibration accelerations according to IRD Mechanalysis [67]

Admissible root-mean-square value of velocity $\tilde{V}$ mm/s	Technical condition of machine: Rigid support	Technical condition of machine: Elastic support
71,0	Unadmissible	Unadmissible
28,0		
18,0		Still
11,2	Still admissible	
7,1		Sufficient
4,5	Sufficient	
2,8		Good
1,8	Good	
1,12		
0,71		
0,45		

Fig. A.4 — Vibration utilization conditions of big turbomachines according to ISO-3945

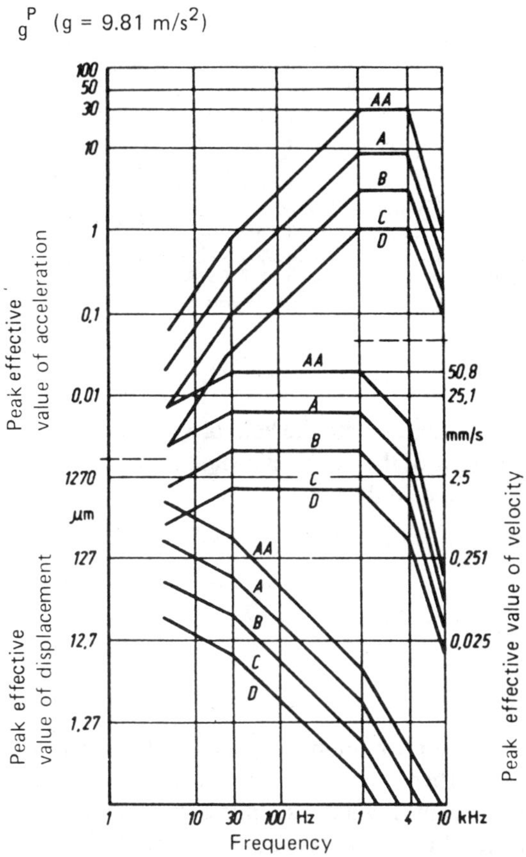

Attention:

AA Catastrophic. Switch off immediately.

A Dangerous. Repair within 48 hrs.

B Partial. Repair within 21 days.

C Not significant. Do not repair.

D Without failures. Do not repair

Note: Root-mean-square amplitude = amplitude measured × repair availability factor *D*. Availability factor *D* for different machine groups:

- electric motor, one stage rotary pump, fan D = 1.0
- typical not critical machines D = 1.0
- turbine, generator, radial compressor, multistage pump D = 2.0
- typical critical machines (due to function) D = 2.0
- centrifuge with rigid shaft, measurement on housing D = 2.0
- centrifuge with suspended shaft D = 0.3
- toothed gear, measurement on the body D = 2–3
- rolling bearings of 30–200 mm diameter D = 0.5
- ferroconcrete foundation D = 2
- different installations with unknown characteristics D = 2

Fig. A.5 — Vibrational classification of machine condition according to Blake [70]. The criterion value is the vibration velocity; the other values are calculated for the dominating vibration frequency

condition and for the calculation of vibration values. The main idea of the part of the slide rule refering to the evaluation of the machine condition and in which the scales are described in the values of effective velocities and acceleration with the consideration of the above-mentioned availability factor *D* is presented in Fig. A.6. It is easy to check on the slide rule that the conversion frequency is $f = 1$ kHz, since 50 mm/s · 2π · 1000 Hz ≅ 316 m/s . It is worth remembering that when we have two measured vibration values, we can calculate the dominating vibration frequency from the formulae for Rice's frequency in Equation (3.50) and compare them with the rotational frequency f_0:

$$f_v = \tilde{a}/(2\pi\tilde{v}) \qquad f_x = \tilde{v}/(2\pi\tilde{x})$$

If, for example, the vibration velocity value claculated only with the consideration of rotations $\tilde{v} = 2\pi f_0 x$ and, with the help of Rice's frequency $\tilde{v} = 2\pi f_x \cdot \tilde{x}$, are close to each other, it means that the only cause of vibrations are the rotations with f_0 frequency. Using this method of reasoning we can increase our knowledge about the behaviour of a machine only having available the vibration meter.

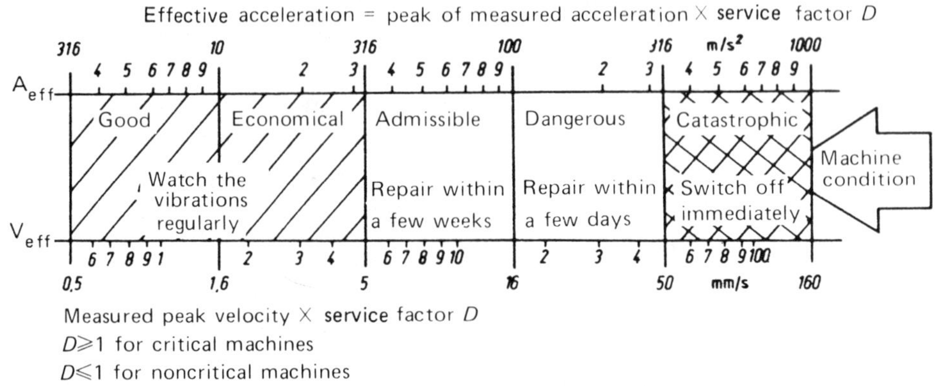

Fig. A.6 — Evaluation of the machine condition with the help of velocities and effective accelerations, as the base of the diagnostic slide rule developed by the Author

APPENDIX 2—DIAGNOSTIC CHART OF ROTATIONAL MOTION MACHINES

As mentioned in the Introduction, experience of vibration diagnostics of machines has been significantly extended since the 1940s. Therefore, it is worth becoming acquainted with the diagnostic chart of rotational machines which supplies the assignment of symptom→ failure relation in the categories of detection probability in a scale from 0 to 10 points (Fig. A.7). Analysing the last row of the chart which has a statistical character, we can arrive at the conclusion that most failures will be detected by measuring the vibrations in the horizontal direction on the shaft or on the bearing housing and for the rotation frequency f_0. Needless to say, the chart below can be used as first approach for designing systems and methods of the diagnostic evaluation of machines. A useful supplement to the chart is formed by the comments about the possible failures of machines.

Table A.1 — Classes of permissible unbalance of rotating rigid machine elements [18]

V_c Peak (mm/s)	Examples of elements subject to balancing
Over 1600	Crankshaft assemblies of low-speed diesel marine engines (rigidly mounted) with an odd number of cylinders
630–1600	Crankshaft assemblies of two-stroke combustion engines rigidly mounted
250–630	Crankshaft assemblies of four-stroke combustion engines rigidly mounted and/or Crankshaft assemblies of diesel marine engines elastically mounted
100–250	Crankshaft assemblies of high-speed four-cylinder diesel engines rigidly mounted
40–100	Crankshaft assemblies of diesel engines with six and more cylinders rigidly mounted. Complete assemblies of automobile engines, truck engines and locomotives
15–40	Automobile and motorcycle wheels, wheel rings, tyres, coupled shafts. Crankshaft assemblies of high-speed combustion engines with four, six (and more) cylinders, rigidly mounted. Crankshaft assemblies of automobile engines, truck engines and locomotives
6.3–16	Coupled shafts with special designation. Rotating parts of crushers and agricultural machines. Crankshaft assemblies of special designation engines with six (and more) cylinders, Crankshaft assembly parts of automobile and truck engines, and locomotives
2.5–6.3	Parts of technological installations. Drums of centrifuges. Fans, flywheels, centrifugal pumps. Parts of machines and machine tools. Rotors of simple electric motors. Crankshaft assembly parts with special designations
1.0–2.5	Rotors of jet engines, steam and gas turbines, turbocompressors, turbogenerators. Drives of machine tools. Rotors of large and medium electric motors with special designation. Rotors of small electric motors. Pumps with turbine drive
0.4–1.0	Drives of record players and tape recorders. Drives of tool grinders and rotors of small electric motors with special designations
Below 0.4	Spindles and discs of precision tool grinders, giroscopes

Fig. A. 7 — Vibroacoustical diagnostics of rotational motion machines—cumulative chart constructed on the basis of IRD Mechanalysis data

Causes of vibrations and noise	Vibroacoustical symptom or symptom probability from 1 to 10	Vibrations																				Noise							
		Maximal amplitude									Dominating frequency (compared to rotation f_0)																		
		Direction			Element																								
		Horizontal	Vertical	Axial	Shaft	Bearings	Body	Foundations	Pipelines	Couplings	0–40% f_0	40–50% f_0	50–100% f_0	$1 \times f_0$	$2 \times f_0$	Higher multipliers	$1/2 f_0$	$1/4 f_0$	Lower multipliers	Odd frequencies	Very high frequencies	Phase (number of points in the stroboscope)	Low-frequency thunder	Booming	Buzzing	Periodic rumble	High squeal	High loud squeak	High screech	Ultrasound
Balancing	Misalignment	5	4	1	9	1								10								(1)		8	2					
	Bent shaft–lost part of rotor	5	4	1	9	1								10								(1)		8	2					
Misalignment	Misalignment	3	2	5	8	1	1							4	5	1						(1) (2) (3)		4	4	2				
Clearances	Clearances	5	4	1		3	2	2	2	1					8	1				1		(2) Roving	8	1	1					
Deformations	Clearances in slide bearing casing	5	4	1	7	1	1			1	1	8	1									(1) Roving	6	2		2				
	Foundation deformation	5	1	4	5	1	1	1		2		2		5	2					1		(1) Roving	1	5	3	1				
	Body deformation	5	4	1	9	1					←1→			8	0.5	0.5						(1) Roving	1	7	1	1				
	Friction of sealing	4	3	3	8	1	1				1	1	1	2	1	1			1	1	1	(1) Roving	2	5		1		1	1	
	Axial friction of rotor	4	3	3	7	1	2				←2→			3	1	1			1	1	1	(1) Roving	3	5		2				

Defects of	Eccentric of pin bearing	5	4	1	9	1							8	2							(1)		1	9					
bearings or pins	Impaired radial bearing	4	3	3	7	2	1				1→		4	2						3	(1) Roving	2	4	1		1	1	1	
	Impaired thrust bearing	3	2	5	6	2	2				9——→									1	(1) Roving	8	1	1					
	Vibration induction in bearing	5	4	1	5	2	2	1			←10→										(1) Roving	6	1		3				
	Unequal rigidity of bearing shaft	5	4	1	4	3	3							9	1						Variable		1	6	2	1			
Defects of	Hysteresis whirl	5	4	1	8	2					8	1	1								(1) Roving	6		2	2				
slide bearings	Oil whirl	5	4	1	8	2						10									(1) Roving	6		1	3				
	Resonance whirl	5	4	1	2	2	2	2	2			10									(1) Roving	6		2	2				
	Frictional whirl	4	3	3	4	2	2	1		1										10	(1) Roving						2	8	
Gear couplings	Impaired gear	5	3	2	8	1	1								2				2	6	(1) Roving	2	1	1	2	2	1		1
	Impaired coupling	4	3	3	7	2				1				1	8	1					(1) Roving			5		5			
Critical work	Critical velocity of shaft	5	4	1	6	4							10								Change by 180°	5	3	2					
conditions	Critical velocity of shaft bearing	5	4	1	7	3							10								Change		5	3	2				
	Critical coupling	4	2	4	1	1				8			10								Change		2	4	2	2			
	Critical support shaft	5	4	1	7	1				2			10								Change		5	4	1				
Resonance	Simple resonance	4	4	2	2	1	2	3	2				10								(1) Roving	4		3	3				
	Subharmonic resonance	3	3	4	2	2	2	2	2									←10→			(1) Jumping	8			2				
	Harmonic resonance	1	4	2	2	1	1	3	3					10										4	2	4			
	Resonance of body	5	4	1		4	4	1	1				8	1		1						2		2	6				
	Resonance of amortisation	5	4	1		2	5	2	1				8	1		1						2		2	6				
	Resonance of foundation	4	3	3		1	4	4	1				8	1		1							1	8	1				
	Torsional resonance	Twist			1	4	4				1		4	2	2				2			1	2	2	3	1	1		

Fig. A. 7 — Vibroacoustical diagnostics of rotational motion machines—cumulative chart constructed on the basis of IRD Mechanalysis data (Continued)

Causes of vibrations and noise / Vibroacoustical symptom or symptom probability from 1 to 10		Vibrations																				Noise							
		Maximal amplitude									Dominating frequency (compared to rotation f_0)																		
		Direction			Element																								
		Horizontal	Vertical	Axial	Shaft	Bearings	Body	Foundations	Pipelines	Couplings	0–40% f_0	40–50% f_0	50–100% f_0	$1 \times f_0$	$2 \times f_0$	Higher multipliers	$1/2\, f_0$	$1/4\, f_0$	Lower multipliers	Odd frequencies	Very high frequencies	Phase (number of points in the stroboscope)	Low-frequency thunder	Booming	Buzzing	Periodic rumble	High squeal	High loud squeak	High screech	Ultrasound
Miscellaneous	Impaired driving belt	4	3	3	5	3		2						◄	10	►						(1) (2) Roving	1	1	3	5				
	Reciprocating forces	3	6	1	5	3	1	1						3	5	2						(1) Roving		8	2					
	Pulsation in pipelines	3	2	5	4	1	1		4					4	5	1						(1) (2)	3	4	3					
	Aerodynamic forces	5	4	1	4	3	2	1						2		6						(1) or more		3	2	1	2	2		
Electric machines	Oval rotor	5	4	1	9	1								10								(1) or more		8	2					
	Misalignment of rotor–stator	4	3	3	8	2								10								Rotating		5	2	3				
	Oval stator	5	4	1	8	2								10								Double point		6	2	2				
	Bent rotor shaft	3	2	5	9	1								10								(1)		8	2					
	Eccentric of electric rotor	3	2	5	6	4								10								(1)		3	3	4				
Average probability of machine failure detection		4.5	3.5	2	6	2	1	0.5	0.3	0.2	0.5	1	0.2	5	1	1	0.2	0.2	0.2	0.2	0.5		2	3	2	2	0.5	0.2	0.1	0.2

Fig. A. 7 — Comments on vibroacoustical diagnostics of rotational motion machines— a cumulative chart

Unbalance	A common cause of vibrations in machines, amplitude propoertional to the magnitude of unbalance. It can cause friction against seals, friction of rotor, damage of bearings. Supporting shafts and rotors can perform relatively great amplitudes of axial vibrations.
Misalignment clearances defromations	Misalignment is revealed by large axial vibrations; if the amplitude of axial component $2f_0$ exceeds 75% of component f_0 radial, then such condition can cause breakdown. Clearances cause many vibration problems, a small clearance can be the cause of dangerous vibrations. Clearances in bearing housings can be confused with oil whirl; they usually accompany unbalance and misalignment. Deformations cause vibrations which indirectly cause misalignment, friction of rotor, uneven load of nearings. Friction is characterized by a very wide spectrum, frequently an ultrasound one; it can cause local heating of rotor (shaft), its bending, as well as cavitation in bearings and resonances.
Defects of bearings or pins	Damaged rolling element bearings generate vibrations with a wide spectrum, particularly in the field of high frequencies. Measurement of accelerations is recommended. The eccentric of shaft pin gets more important as the vibration source when the shaft is terminated with a gear.
Defects of journal bearings	*Hysteresis whirl* Rare but dangerous. It is revealed by the whirling of the shaft pin in the direction of rotation. It is evoked by the destabilizing action of the forces of internal damping (hysteresis damping) of the shaft. These forces are displaced in phase by 90° with respect to the shaft bending stress. *Oil whirl* In the direction of shaft rotation, it is evoked by the pressure wave in the layer of slide bearing oil. Vibration induction can take place around 1/2 f_0. Similar causes are given by the resonance vortex. *Friction whirl* In opposite direction to shaft rotation, it is created as a result of the instantaneous direct and dry contact of shaft with the bearing bushing. Original cause: faulty lubrication, extensive movement of pin in the bearing bushing.
Coupling gears	Misalignment and eccentricity are the original causes of gear damage; pitting and pinholes are the result of uneven load distribution. Couplings are sensitive to misalignment and torsional vibrations. Vortices in slide bearings can also be dangerous.
Critical work conditions	The essential differences between the critical work conditions and the resonance conditions of any optional element consists of the absence of change in the direction or type of stress, e.g. the shaft in

the critical condition does not oscillate but whirls with a great deflection without any change in the bending stress direction. The critical condition can be removed by the balancing of shafts, rotors, couplings, etc., or the change of rotational frequency f_0.

Resonance Resonance does not generate but intensifies the vibrations from other sources. It can cause a dangerous situation, intensifying the rated vibrations of machines or pulsations in pipelines. It is frequently the cause of rotational vibrations in journal bearings. Torsional vibrations are usually not noticed since this motion is superimposed on shaft revolution; hence, damage can appear unexpectedly. If a gearbox produces torsional vibrations, they can be revealed by increased noise.

Miscellaneous Impaired driving belt is quite visible with a stroboscope; after replacement care should be taken to get even belt tension and alignment of shafts. The to-and-fro forces unseparable with crankshaft machines can be decreased by constructional changes or by vibroisolation. Aerohydrodynamic forces usually have a frequency (e.g. number of baldes $\times f_0$); these pulsations can produce resonance vibrations.

Electric machines Electrical causes occur with a frequency 1 or 2 × the frequency of the supply current and they disappear when the current is switched off. Mechanical causes occur as modulation factors of resonance vibrations of stator and rotor, etc.

Noise radiation Mechanical and electrical defects are the sources of vibrations and noise. Mechanical nosie can follow from unbalance, bearing vibrations, misalignment and vibration of other associated construction elements.

Noise of electrical origin occurs due to the transformation of electric energy by: (a) variable magnetic forces as the function of the density of induction flux, the number and form of poles and geometry of air gap; (b) random phenomena, sparking of bushes, etc.

Aerodynamic noise follows from the detachment of vortices around stationary and movable obstacles, pulsation of medium and mixing of gases with different velocities; it can have the wide-band or narrow- band character.

Wide-band nosie: (a) blades of fluid-flow machines, obstacles, supporting constructions in the fluid stream, driving belts, etc.; (b) sudden change in the direction or cross-section of the stream, mixing of streams, etc.

Narrow-band noise: (a) resonance effect of organ pipe, vibrating cables, covers or construction elements in airstream; (b) edge tones on sharp edges, air columns induced by flow; (c) siren effect on rotating machine elements.

Impact noise created by direct collision of elements, e.g. by a falling

Fig. A. 7 — Comments on vibroacoustical diagnostics of rotational motion machines— a cumulative chart (Continued)

---

hammer, pneumatic hammer. Impairment of teeth in the gear can be heard or be silent; similarly the clapping of a faulty driving belt. These impacts can follow so quickly one after another that in order to detect them a special technique of quick recording is necessary. Areas with many sources of impacts can emit permanent ringing as a result of superimposing many impacts.

APPENDIX 3—CRITERIA OF ROTOR BALANCE

In Section 2.2.1 we discussed unbalance as the reason for many vibration problems in the use of machines. The criterion value in the balancing, as we indicated there, is the velocity v_c of the motion of machine element due to rotations ω_0 on the radius e of its unbalance, since $v_c = e\omega_0 = 2\pi e \cdot f_0$. Table A.1 presents the classes of admissible unbalance as the limits of velocity v_c for different machine elements.

From the values presented in the table of admissible velocities of vibrations we can estimate the vibration velocity of the whole machine, since the velocity v_c is transferred in the form of the equivalent inertia force through the bearings on the machine body. When we determine the weight of the rotor by m_w, that of the body by m_k, and the velocity of body vibration by v_k, then we can write the following equation of the balance of inertia forces after mounting in the bearings:

$$m_w \omega_0 v_c = \left(m_w + m_k\right) \omega_0 \cdot v_k$$

hence, the estimation of vibrationvelocity of the body with $m_k = (1 + 3)\, m_w$:

$$v_k = \frac{m_w \cdot v_c}{m_w + m_k} = \left(\frac{1}{2} + \frac{1}{4}\right) v_c \Rightarrow x_k = \left(\frac{1}{2} + \frac{1}{4}\right) x_c$$

If, in the light of this result, we take into consideration the parts of technological installations $v_c = 2.5–6.3$ mm/s (Table A.1) then the velocity of body vibrations measured on the bearings with $m_w = m_k$ will be $v_k = 1.25 + 3.15$ mm/s, which is a commonly measured value.

Table A.1 does not specify the velocities of the rotational elements, i.e. the admissible forces which can be transferred to the bearing. A specification of these data is presented in Fig. A.8 which shows the proper balancing e in μm or mm, and the unbalancing $N = e \cdot m_w$ in grams,millimetres per kilogram or kilograms, millimetres per kilogram for the given rotation velocity of the rotor. For the constant admissible velocity v_c the admissible unbalance decreases inversely proportionally to the rotation velocity.

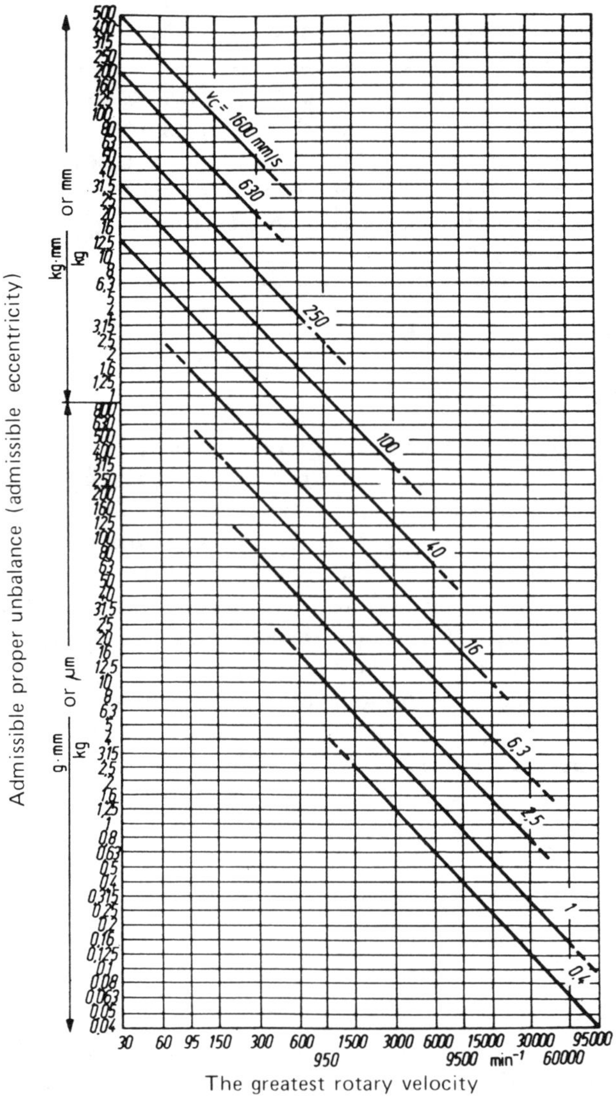

Fig. A.8 — Admissible unbalance of rotational machine elements as a function of the unbalance class V_c and rotation velocity according to ISO-1940

APPENDIX 4—COMPUTATION NOMOGRAM OF HARMONIC VIBRATIONS

In many cases of diagnostics we know the vibration frequency (e.g. the rotational frequency f_0) and the amplitude of one vibration value, e.g. the measured value of vibration displacement. However, we would like to know the amplitudes of the remaining vibration values, even with the simplest assumption of harmonic vibrations. We know that:

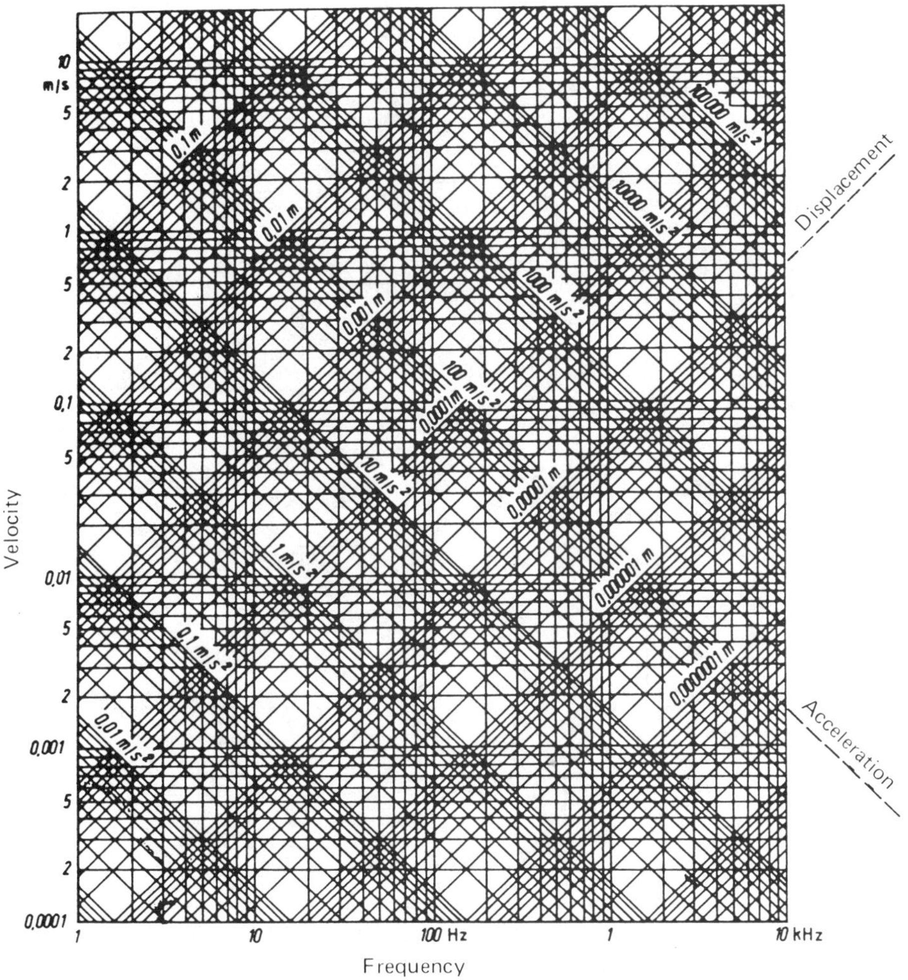

Fig. A.9 — Nomogram of acceleration, velocity, displacement for harmonic vibrations with a determined frequency

$$a = 2\pi \cdot f \cdot v = 4\pi^2 \cdot f^2 \cdot x$$

which is the basis of the nomogram presented in Fig. A.9. This nomogram can be used not only for the rotation frequency of vibrations but particularly for the dominating frequency, i.e. for Rice's frequencies f_v, f_x measured instrumentally or calculated according to the Equation (3.50) (see Appendix 1).

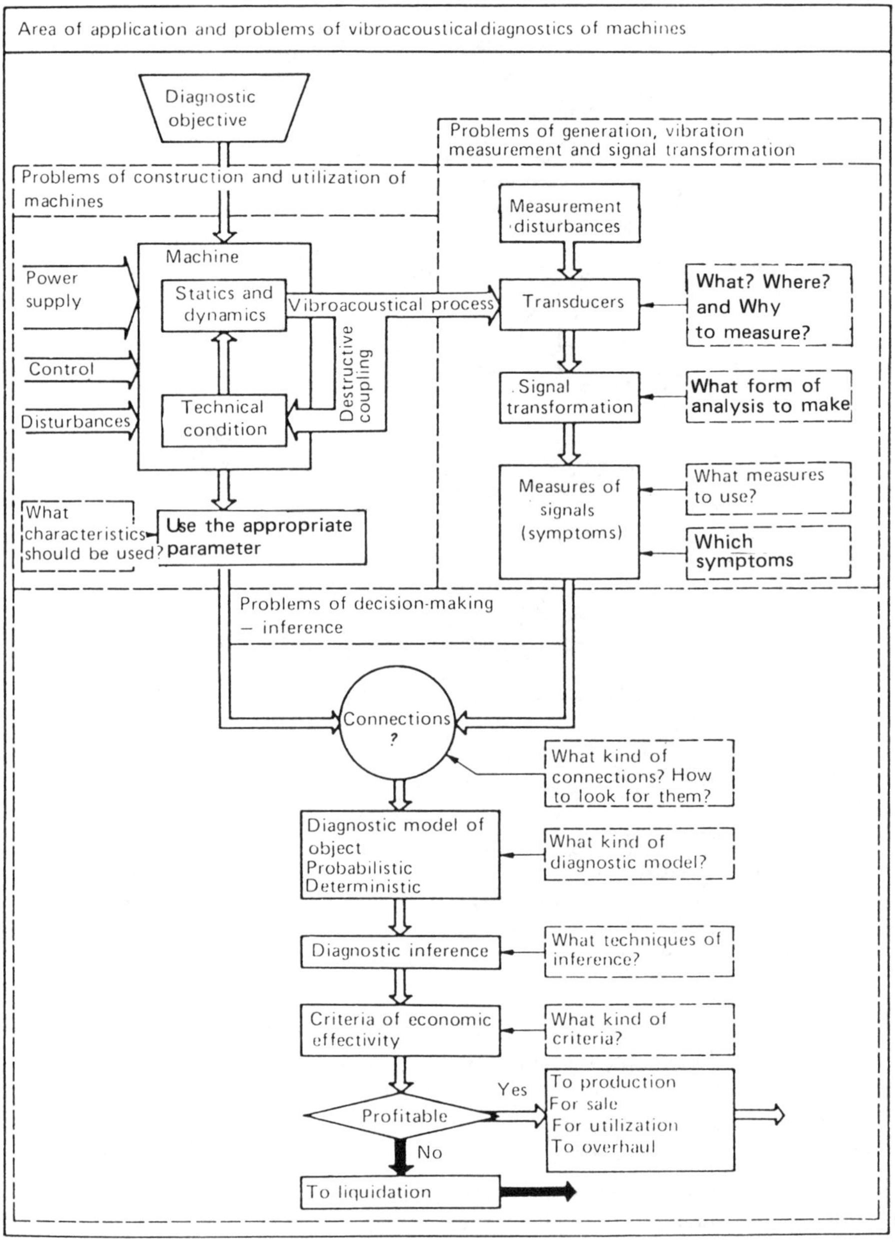

Fig. A.10 — Diagram of the area of vibroacoustical diagnostics application and its problems

Conclusion

It is appropriate to conclude this book with an analysis of diagram (Fig. A.10) presenting the total problem of the machine vibration diagnostics. In order to be successful to be successful in the creation and application of diagnostic methods, we should have an understanding of the problems of construction, manufacturing, and the use of the diagnosed objects. These problems are discussed in Chapters 1 and 2. Chapter 2 gives the bases of the generation of tribovibroacoustical phenomena and without knowing them it is impossible to make a selection and decision regarding the questions: What; Where; and: How to measure?, according to Fig. A.10. Chapter 3 gives an answer to the question referring to the processing of signals, the selection of measures and symptoms colinear with the investigated mode of machine wear. Chapter 4 answers the question referring to inference about the present and future condition of the diagnosed machine. More recent methods referring to the problems of the processing of signals and symptoms (i.e. in general terms the diagnostic information) are presented in Chapter 6 which shows already confirmed conceptions or those being under development and aiming at the improvement of the relation between signal and noise.

The vibroacoustical problems of machine diagnostics presented in this way seem to be coherent and complete in the sense that each problem has at least one solution method. The state of knowledge in the field of diagnostics is at the beginning of its development, which means that many problems are still far from having more than one solution. Progress will be achieved with the natural effect of the accumulation of mature considerations of experimental facts and trials of theoretical explanation. It is an illustration of the known fact that first we have an invention and then follow its scientific justifications. This state should be an incentive for wide implementations of vibroacoustical diagnostics. Feedback will lead to the extension of the diagnostic knowledge. The fulfilment of both these objectives is hopefully expected by the Author and other researchers in this growing field of engineering.

References

[1] Ziemski, S., 1973. *Problems of good diagnosis* (in Polish). Wiedza Powszechna, Warsaw, p.17.

[2] Pszczołowski, T. 1978. *Small encyclopedia of praxeology* (in Polish). Ossolineum, Wrocław, p.48.

[3] Collacott, R.A., 1977. *Mechanical fault diagnosis.* Chapman & Hall, London, Chapter 11.

[4] Lewitowicz, J., 1983. 'Physicochemical bases of maintaining machines in operation' (in Polish). *Materials of a conversatorium devoted to the problems of maintaining machines in operation,* Wisła, 1983, p.97–118.

[5] Dietrych, J., 1978. *System and construction* (in Polish). WNT, Warsaw, p.31.

[6] Cempel, C., 1982. *Mechanical vibrations* (in Polish). Poznań University of Technology, Chapter 1.

[7] Cempel, C., 1982. *Foundations of vibroacoustical diagnostics of machines* (in Polish). WNT, Warsaw.

[8] Bedkowski, L., 1981. 'General theory of technical diagnostics' (in Polish). Habilitation dissertation. *WAT Bulletin No. 3* (supplement), p.25.

[9] Neale, M.J. and Woodley, B.J., 1979. 'Condition monitoring methods and economics.' *Brüel-Kjaer Application Note,* Denmark Naerum.

[10] Hudachek, R.J. and Dodd, V.R., 1976. 'Refinery machinery surveillance and diagnostic program pays off.' *Oil and Gas Journal,* October 18, pp.70–82.

[11] VCI (Vibration Consultants Limited) England, 1980. Advertising materials.

[12] Birger, I.J., 1978. *Technical diagnostics* (in Russian). Mashinostroyenye, Moscow, p.12.

[13] 'Machine maintenance—the role and vibration measurement and analysis. *Brüel-Kjaer Technical Report 017-80.*

[14] Bobrowski, D., 1977. *Mathematical introduction to the reliability theory* (in Polish). Poznań University of Technology, p.26.

[15] Miroshnikov, L.W., Noljdin, A.P. and Pal, W.I., 1977. *Diagnosing the technical condition of cars in car transport enterprises.* Transport, Moscow. (in Russian).

[16] Kožesnik, J., 1963. *Dynamics of machines* (in Polish). WNT, Warsaw, Chapter 3.

[17] Leyko, J., 1969. *General mechanics* (in Polish). PAN, Warsaw, Vol. 2. Chapter 11.

[18] Laczkowski, R., 1979. *Balancing of rotary elements* (in Polish). WNT, Warsaw, Section 2.2.

[19] Laczkowski, R., 1983. *Vibroacoustics of machines* (in Polish). WNT, Warsaw, Section 2.1.4.

[20] Nicolet Scientific Corporation, USA, 1981. *Machinery vibration diagnostic guide for Mini Analyzer 100 A.*

[21] Raguliskis, K.M., *et al.*, 1976. *Vibrations of turbo-systems* (in Russian). MOKSLFS, Vilna, Chapter 3.

[22] Cempel, C., 1970. *Periodical vibrations with impacts in discrete mechanical systems* (in Polish). Poznań University of Technology, Dissertation series No. 44, Section 2.1.

[23] Den Hartog, J.P., 1971. *Mechanical vibrations* (in Polish). PWN, Warsaw, Section 6.6.

[24] Rothbart, H.A., (Ed.), 1964. *Mechanical design and system handbook.* McGraw-Hill, New York, Section 12.7.

[25] Harris, C.M. and Crede, Ch.E. (Eds), 1976. *Shock and vibration handbook,* 2nd ed, McGraw-Hill, New York, Chapter 5.

[26] Collective work, 1980. *Diagnostics of rolling bearings* (in Polish). Ossolineum, Wroclaw, pp.9-41.

[27] Shubov, I.G., 1974. *Noise and vibration of electric machines* (in Russian). Energia, Leningrad, Chapter 4.

[28] Dawson, D. and Taylor, C.N., 1979. 'Cavitation in bearings.' *Annual Review of Fluid Mechanics,* No. 11, pp.35–66.

[29] Luczak, A. and Mazur,T., 1981. *Physical aging of machine elements* (in Polish). WNT, Warsaw, Section 3.4.

[30] Ross, D., 1976. *Mechanics of underwater noise. Pergamon Press, New York, Chapter 7.*

[31] Moore, D.F., 1975. *Principles and application of tribology.* Pergamon Press, Oxford, Chapter 3. (Russian translation MIR, 1978.)

[32] Cempel, C., 1980. *Acoustical emission and its utilization* (in Polish). *Eksploatacja i Dozór,* No. 7/1980, pp.22–25.

[33] Ivanov, W.I., 1980. 'Adaptation of acoustical emission method to non-destructive inspection and examination of materials' (in Russian). *Defectoscopia,* No. 5, pp.65–84.

[34] Tverskoi, W.S., 1981. 'Engineering analysis of the spectra of acoustical emissions' (in Russian). *Defektoskopia,* No. 5, pp.62–67.

[35] Wnuk, M.P., 1977. *Bases of the mechanics of cracking* (in Polish). AGH, Cracow, Chapter 3.

[36] Dunagan, H.L., 1975. 'Using acoustical emission technology to predict structural failure.' *Metals Engineering Quarterly,* February 1975 (reprint).

[37] Buch, A., 1964. *Problems of fatigue resistance* (in Polish). PAN, Warsaw, Chapter 1.

[38] Juvinall, R.C., 1967. *Engineering considerations of stress, strain.* McGraw-Hill, New York, Chapter 11.

[39] Cempel, C., 1983. 'The fatigue limits for vibration of machine and structural elements.' *Zagadnienia Eksploatacji Maszyn,* No. 4, pp.51–61.

[40] Kocańda, S., 1978. *Fatigue destruction of metals* (in Polish). WNT, Warsaw, p.278.

[41] Crandal, S.H. (Ed.), 1963. *Random vibration,* Vol.2, Chapter 5. MIT Press. (Russian edition).

[42] Tshatinian, R.M., 1966. Energetic criteria of fatigue resistance in machine details

(in Russian). *Vestnik Mashinostroyenia*, No. 8, pp.27–29.

[43] Pronikov, A.S., 1978. *Reliability of machines* (in Russian). Mashinostroyenie, Moscow, Chapter 5.

[44] Engel, P.A., 1976. *Impact wear.* Elsevier, Amsterdam, Chapter 1.

[45] *I Symposium on Material Creeping,* Technical University of Bialystok, 22–24 September 1983. Particularly: Juskowlul, A., 'Mechanisms of creeping and vibro-creeping' (in Polish), pp.81–93

[46] Ultrasound, 1979. *Small encyclopedia* (in Russian). Izdatelstvo Soviet Encyclopedia, Moscow, p.250.

[47] Astashev, W.K., 1983. Influence of high-frequency vibration on the processes of plastic deformation (in Russian). *Mashinostroyenie,* No. 2, pp.3–11.

[48] Cempel, C., 1980. Vibroacoustical signal as the measure of tribological process (in Polish). *Materials of the 10th Tribological School,* September 1980. WITPIS, pp.194–203.

[49] Cempel, C., 1988. Vibroacoustical diagnostics of machinery—an outline. *Mechanical Systems and Signal Processing,* Vol. 2 (2), pp.135–151.

[50] Cempel, C., 1976. *Synchronic summarization as a method of machine diagnostics* (in Polish). *Arch. Bud. Maszyn,* Vol. 23, No. 3, pp.433–441.

[51] Cempel, C., Smarsz, T. and Szafranek, A., 1980. Evaluation of the technical condition of a combustion engine with the help of the estimates of pulsation signal (in Polish) *Second Conference on Vehicle Diagnostics,* Pila-Tuczno, 1980, pp.25–44.

[52] Cempel, C., Smarsz, T. and Szafranek, A., 1980. Evaluation of the technical condition of a combustion engine with the help of spectrum (in Polish) *Second Conference on Vehicle Diagnostics,* Pila-Tuczno, 1980, pp.45–59.

[53] Bendat ,J.S. and Piersol, A.G., 1971. *Random data: analysis and measurement procedures.* Wiley Interscience, New York, Chapter 5 (Polish edition).

[54] Tomović, R. and Vukobtratovitsh, M., 1972. *General sensitivity theory* (in Russian). Sovietskoye Radio, Moscow, Chapter 1.

[55] Kawalerow, G.I. and Mandelsztom, S.M., 1983. *Introduction to the theory of measurements* (in Polish). PWN, Warsaw, Chapter 1 (translation from Russian).

[56] Sobczyk, K., 1973. *Methods of statistical dynamics* (in Polish). PWN, Warsaw, Chapter 25.

[57] Gersch, W., *et al.,* 1980. Nearest neighbour time series analysis classification of faults in rotating machinery. Reliability, stress analysis and failure prevention methods in mechanical design. *ASME Conference 1980,* pp.51–58.

[58] Rimski-Korsakov, A.W., 1973. *Electroacoustics.* (in Russian). Izdatielstvo Svyaz, Moscow, Section 4.3.

[59] Huton, P.H. and Skorpik, J.R. 'In flight fatigue crack monitoring using acoustic emission.' *ISA Transection,* Vol. 20, No. 2, pp.79–83.

[60] Bracewell, R., 1968. *Fourier's transform and its application* (in Polish). WNT, Warsaw, Chapter 6.

[61] Balderstone, H.L., 1969. Detection of incipient failure in bearings. *Materials Evaluation,* June, pp.121–28.

[62] Szabatin, J., 1982. *Bases of signal theory* (in Polish). WKŁ, Warsaw, Chapter 8.

[63] Pavlov, B.W., 1971. *Acoustical diagnostics of mechanisms* (in Russian). Mashinostroyenie, Moscow.

[64] Mitchell, J.S. 'Examination of pump cavitation, gear mesh and blade performance using external vibration characteristics'. *Proceedings Fourth Turbomachinery Symposium,* University of Texas, pp.39–45.

[65] Deutch, R., 1969. *Theory of estimation* (in Polish). Warsaw, p.64.

[66] Dodds, C.J. and Robson, J.D., 1975. 'Partial coherence in multivariate random processes.' *Journal of Sound and Vibration,* Vol. 42, No. 3, pp.243–49.

[67] Special treatment of vibration sources to reduce plant noise. *IRD Mechanalysis Technical Paper No, 110,* 1976.

[68] *IRD Mechanalysis.* Prospects and materials. Ohio, 1975.

[69] 'Determining vibration warning levels.' *Application Note No. 803.* PMC/BETA Corporation, USA.

[70] Blake, M.P. and Mitchel, W.S., 1972. *Vibration and acoustic measurement handbook.* New York, Spartan Books, Chapter 22.

[71] Wojciechowski, R., 1979. *Utilization diagnostics of rolling element bearings in electric motors.* Master's thesis IMT, Poznań University.

[72] Andrzejewski, M. and Gospoś, W., 1983. 'Analysis of the technical condition of a pumping engine' (in Polish). *Eksploatacja Maszyn,* No, 7, pp.11–13.

[73] Andrzejewski, M. *et al.,* 1984. 'Vibration diagnostics of a high-power electric motor' (in polish). *Eksploatacja Maszyn,* No, 5/6, pp.17–18.

[74] Dyer, D. and Stewart, R.M., *Detection of rolling element bearing damage by statistical vibration analysis.* American Society of Mechanical Engineers Paper No. 77-DET-83

[75] Rogers, L.M., 1979. 'The application of vibration analysis and AE source location to on line condition monitoring of antifriction bearings.' *Tribology International,* Vol. 12, No. 2, pp.51–59.

[76] Cempel, C., 1980. Diagnostically oriented measures of vibroacoustic processes. *Journal of Sound and Vibration,* Vol. 73, No. 4, pp. 547–61.

[77] Papoulis, A., 1972. *Probability, random variables and stochastic processes* (in Polish). WNT, Warsaw, Chapter 10.

[78] Bannister, R.L. and Donato, V., 1971. Signature analysis of turbomachinery. *Sound and Vibration,* No. 9.

[79] Randall, R.B. Cepstrum analysis and gearbox fault diagnosis. *Brüel-Kjaer Application Notes,* pp.13–150 and 233–80.

[80] Fijalkowska, J., 1984. *Cepstrum and product spectrum in the diagnostics of turbomachines* (in Polish) Master's thesis IMT Poznań Technical University.

[81] Golec, M., 1981. *Evaluation of the usability of discriminants of VA processes in machine diagnostics* (in Polish). Doctor's dissertation. Poznań Technical University.

[82] Golec, M. and Cempel, C. Evaluation of the utility of the spectral measures of VA processes in machine diagnostic' (in Polish). *Zagadnienia Eksploatacji Maszyn* (in print).

[83] Cempel, C., 1982. Vibroacoustical diagnostics of machines—survey of the problems (in Polish), *Sixth Symposium of Vibration Technics and Vibroacoustics,* 9–11 December 1982. AGH 1984, pp.33–50.

[84] Tou, J.T. and Gonzales, R.C., 1974. *Pattern recognition principles.* Addison-Wesley, London (Russian edition).

[85] Cholewa, W., 1983. Method of machine diagnosing with the application of fuzzy sets (in Polish). *Scientific Papers of Silesian Technical University,* No. 764,

Supplement A. Gliwice. Habilitation dissertation.
[86] Batchelor, G. (Ed.), 1978. *Pattern recognition.* New York, Plenum Press. Chapter 4.
[87] Piasecki, S. Evaluation of the utilization quality of machines and vehicles (in Polish). *Fifth Congress on Utilization Materials,* pp.50–56.
[88] Volk, V., 1973 *Applied statistics for engineers* (in Polish). WNT, Warsaw, Chapter 10.
[89] Cempel, C., 1984. The vibration symptom limit value in condition monitoring. *Condition Monitoring Conference Swansea 1984,* pp.328–39.
[90] De Groot, M.H., 1981. *Optimal statistical decisions* (in Polish). PWN, Warsaw.
[91] Cempel, C., 1984. Accuracy of vibroacoustical diagnosis of mechanical objects with a multi-dimensional vector of symptoms (in Polish). *Zagadnienia Eksploatacji Masyzn,* No. 1–2, pp.253–57.
[92] Cempel, C., 1983. Determination of the limit value of symptom in VA machine diagnostics (in Polish). *Materials of the Fifth Congress of Utilization,* Warsaw 1983, pp.2010–2020.
[93] Krugiełka, C. C., 1983. *Vibroacoustical turbo-machine diagnostics on the example of the PAK power station complex.* Master's thesis. Institute of Technical Mechanics, Poznań Technical University.
[94] Cempel, C., 1984. Determination of the limit condition in vibrational diagnostics of the auxiliary machines of power unit. *Energetyka,* (in print).
[95] Nowicki, R. and Sordyl, F., 1984. 'Determination of the limits of utilization condition in diagnostic examinations' (in Polish). *LDH Report No. 138.* Poznań Technical University.
[96] Cempel, C., 1984. Diagnostic differentiation of the condition of mechanical objects in the presence of disturbances (in Polish). *Zagadnienia Eksploatacji Masyzn,* (in print).
[97] Krzystofiak, M. and Urbanek, D., 1978. *Statistical methods* (in Polish). PWN, Warsaw, p.363.
[98] Box, G.E.P. and Jenkins, G.M., 1983. *Analysis of time series* (in Polish). PWN, Warsaw, Section 3.3.
[99] Zelias, A., 1979. *Theory of prognosing* (in Polish). PWN,Warsaw, Section 3.6.
[100] Batko, W., 1984. *Methods of a synthesis of prediction diagnoses in technical diagnostics* (in Polish). Habilitation dissertation, AGH, Cracow, 1984.
[101 Tomaszewski, F., 1981. *An attempt of vibroacoustical diagnostics of a locomotive combustion engine and a reliability and economic analysis* (in Polish). Diploma work of Postgraduate Studies in Vibroacoustics of Machines, Poznań Technical University.
[102] Tomaszewski, F., 1983. 'Diagnostics of diesel engines' (on Polish). *Eksploatacja Maszyn,* No. 5/6, pp.17–18.
[103] Porwich, C., 1984. *Vibrational evaluation of textile machines condition on the example of the driving units of godets of TORAY Co.* (in Polish). Postgraduation Studies in Vibrational Diagnostics of Machines. Diploma work. Poznań Technical University.
[104] Kowalak, J. and Golec, M., 1981. 'Diagnostics of rolling element bearings in electric motors with the utilization of the estimators of vibration signal' (in Polish). *Seventh Symposium of Machine Diagnostics,* Wisla, pp. 50–60.
[105] Collective work, 1980. 'Diagnostics of rolling bearings' (in Polish). *IV School of*

Diagnostics. Ossolineum, Wrocław. pp. 8–240.
[106] Kowalak, J. and Tokarz, K., 1984. Influence of technological and assembly errors on the displacement of the external ring of roller bearings. (in Polish). *Mechanik*, 2, pp.93–94.
[107] Collective work, 1981. 'Diagnostics of vehicles' (in Polish). *V School of Diagnostics*. Ossolineum, Wrocław. pp. 7–301.
[108] Collective work, 1983. Computer transformation of diagnostics signals (in Polish). *VI School of Diagnostics*. Poznań Technical University, Poznań-Rydzyna.
[109] Niziński, S., Mazur, T. and Pelc, H., 1976/ Acoustic examination of the technical condition of engines 1160-076 (in Polish). *Second Symposium of Machine Diagnostic*, Porebka, pp.116–41.
[110] Hebda, M., Niziński, S. and Strzelecki, A., 1977. 'Identification of the condition of a piston and crankshaft mechanism of FIAT-125 engine by the method of vibration measurement' (in Polish). *III Symposium of Machine Diagnostics*, Szczyrk, pp.65–91.
[111] Krzyworzeka, P., 1972. *Optimization of the processing of a complex diagnostic signal* (in Polish). Doctor's dissertation, AGH, Cracow.
[112] Adamczyk, J., 1979. *Selected problems of vibroacoustical machine diagnostics* (in Polish). Habilitation dissertation. ZN AGH, Cracow, Electrification and Mechanization of Mining and Metallurgy. Fascicle 101.
[113] Anderson, T.W., 1958. *An introduction to multivariate statistical analysis*. Chapman and Hall, London, Chapter 12 (Russian translation.)
[114] Żółtowski, B., 1978. *Evaluation of the utility of VA processes in the diagnostics of rear axle drives on the example of the car* BRDM-2 (in Polish). Doctor's dissertation. Piła WOSzS.
[115] Sobczak, W. and Malina, W., 1978. *Methods of information selection* (in Polish). WNT, Warsaw.
[116] Krzyśko, M., 1982. *Discrimination analysis* (in Polish). Edition of Adam Mickiewicz University, Poznań.
[117] Cempel, C., 1980. Reduction of data set in machine diagnostics (in Polish). *Zagadnienia Eksploatacji Maszyn*, No. 4, pp.571–84.
[118] Ahrens, H. and Läuter, J., 1979. *Multidimensional analysis of variance* (in Polish). PWN, Warsaw, Section 4.4.
[119] Deutsch, R., 1969. *Theory of estimation* (in Polish). PWN, Warsaw, Section 3.3.
[120] Korn, G.A. and Korn, T.M., 1968. *Mathematical Handbook* (2nd edn). McGraw-Hill, New York (Russian edition).
[121] Kowalski, K., 1982. Determination of VA signal evaluation by the method of orthogonal filtration and moments of higher orders (in Polish). *LDH Report No. 108*, Poznań Technical University.
[122] Tadeusiewicz, R., 1983. Methods of image identification in diagnostics (in Polish). Computer Transformation of Signals. *Sixth School of Diagnostics*, Rydzyna, pp.215–60.
[123] Widrow, B., *et al.*, 1975. Adaptive noise cancelling: principles and applications. *IEEE Proceedings*, Vol. 63, No. 12, pp.1692–716.
[124] Chaturvedi, G.K. and Thomas D.W., 1981. Adaptive noise cancelling and condition monitoring. *Journal of Sound and Vibration*, Vol. 76, No. 3, pp.391–405.
[125] Chaturvedi, G.K. and Thomas D.W. Bearing fault detection using adaptive noise

cancelling. Trans. American Society of Mechanical Engineers. *Journal of Mechanical Design,* Paper No. 81-DET-7, pp.7–10.

[126] Cempel, C., 1985. Discrimination analysis and designing of filters in VA machine diagnostics (in Polish). *Seventh School of Diagnostics,* Rydzyna, pp.57–69.

[127] Hebda, M., Niziński, S. and Pelc, H., 1980. *Bases of the diagnostics of mechanical vehicles* (in Polish). WKŁ, Warsaw.

[128] Collective work, 1983. Computer transformation of diagnostic signals (in Polish). *VI School of Diagnostics,* Poznań Technical University, Poznań-Rydzyna.

[129] Mitchel, J.S., 1981. *Machinery analysis and monitoring.* Penn Well Publishing Company, Tulsa, Oklahoma.

[130] Hebda, M. and Janicki, D., 1977. *Durability and reliability of cars in utilization* (in Polish). WKŁ, Warsaw.

Index